# Synthesis Lectures on Renewable Energy Technologies

The series, Synthesis Lectures on Renewable Energy Technologies publishes concise books, focused on technologies that harness energy from naturally occurring sources, such as sunlight, wind, water, geothermal heat, and biofuels from organic materials. These renewable energy technologies play a crucial role in transitioning away from fossil fuels, helping to mitigate the effects of climate change, and promoting a sustainable energy supply.

Luofeng Huang

# Ocean Renewable Energy Systems

## Harnessing Wind, Wave, Tidal, and Solar Power at Sea

Springer

Luofeng Huang
Faculty of Engineering and Applied Sciences
Cranfield University
Cranfield, UK

ISSN 2690-5000 ISSN 2690-5019 (electronic)
Synthesis Lectures on Renewable Energy Technologies
ISBN 978-3-032-10575-2 ISBN 978-3-032-10576-9 (eBook)
https://doi.org/10.1007/978-3-032-10576-9



This Springer imprint is published by the registered company Springer Nature Switzerland AG
The registered company address is: Gewerbestrasse 11, 6330 Cham, Switzerland

*To Hanna Räsänen and River Huang*

**Acknowledgements** I am deeply grateful to Arza Seidel, whose exceptional editorial guidance has been invaluable; this book would not have been possible without her insight and dedication. My sincere thanks also go to the Springer Nature team, in particular Prasanna Kumar Narayanasamy, for their thoughtful support in shaping this work into its best form.

I would also like to acknowledge the unwavering encouragement of my colleagues. In particular, I am indebted to Dr. Ali Nabavi, Prof. Phil Longhurst, Ms. Louise Wootton, Prof. Chris Fogwill, Prof. Phil Hart, Prof. Joy Sumner, Prof. Jonas Ringsberg, Prof. Tahsin Tezdogan, Prof. Giles Thomas, Prof. Ying Xie, Dr. Jerry Luo, Dr. Qing Qin, Dr. Yifeng Yang, Dr. Danlei Yang, Dr. Hashim Elzaabalawy, Dr. Patrick Verdin, Mr. Tim Wood, Mr. Chenhao Mi and Mr. Binjian Ou, whose support and camaraderie have been truly invaluable throughout this journey.

**Competing Interests** The author has no competing interests to declare that are relevant to the content of this manuscript.

# Contents

# Introduction 1

## 1.1 Evolvement of Global Energy and Offshore Opportunities

For much of modern history, the global energy system has been dominated by fossil fuels—coal, oil, and natural gas. These energy sources powered the Industrial Revolution, drove economic growth throughout the nineteenth and twentieth centuries, and continue to account for the majority of the world's primary energy consumption. In 1990, the world's energy consumption was over 80% from fossil fuels, alongside around 5% of nuclear and 5% of hydropower. While wind and solar's contribution was negligible at the time. Since then, the energy landscape has begun a transformation. Renewable energy, especially wind and solar energies, has grown rapidly over the last three decades. By 2024, wind and solar have reached a combined share of more than 5%, whilst fossil fuels' share has been reduced to around 75% [1].

Despite rapid growth, the shift to renewables is not just a strategic choice. It is an urgent imperative. Global warming has reached approximately 1.2 °C above pre-industrial levels [2]. Sea levels have subsequently risen by over 20 cm since 1900, and the pace of increase is accelerating due to both ocean warming and ice-sheet loss in polar regions, placing people in low-lying coastal zones at heightened risk [3]. Observed impacts—including record-breaking heatwaves, floods, and hurricanes—have already generated over $4 trillion in global economic losses between 1970–2021 [4].

One proposed strategy to mitigate emissions from fossil fuel use is Carbon Capture and Storage (CCS)—a process that captures $CO_2$ emissions and stores them in enclosed places. A related approach is Direct Air Capture (DAC), which removes $CO_2$ directly

The original version of the chapter has been revised. A correction to this chapter can be found at https://doi.org/10.1007/978-3-032-10576-9_8



L. Huang, *Ocean Renewable Energy Systems*, Synthesis Lectures on Renewable Energy Technologies, https://doi.org/10.1007/978-3-032-10576-9_1

from ambient air, often requiring large-scale fans and chemical solvents. While technically feasible, CCS remain costly, energy-intensive, and slowhard to scale. Currently, CCS operations combined capture only around 0.1% of global annual $CO_2$ emissions [5]. This underscores the limitations of relying on carbon removal technologies to meet climate targets. Instead, preventing emissions at the source through rapid deployment of renewables is widely recognized as the most effective path to deep decarbonisation.

The contemporary transformation of fossil fuels into renewable energies is just the beginning. According to the International Energy Agency (IEA)'s Net Zero by 2050 scenario, wind and solar are projected to supply nearly 70% of global electricity generation by mid-century, with solar PhotoVoltaic(PV) expected to become the single largest source of energy, as shown in Fig. 1.1. By 2050, fossil fuels' share is predicted to drop to a minor level, mostly supported by carbon capture or reserved for non-energy uses. This dramatic shift marks a fundamental transformation of the global energy system, positioning wind and solar at the core of a cleaner, more sustainable energy future [6].

Today, wind and solar are leading the growth in renewable energy. Global installed capacity of solar PV rose from just 40 GW in 2010 to more than 1400 GW by 2023, while wind energy rose from 180 GW to over 1000 GW in the same period. With the development of renewable technologies, the cost of solar modules dropped by over 90%, and wind turbine prices fell by around 70%, making both technologies highly competitive with fossil fuels in many regions. Levelized Cost of Energy (LCOE) is the average cost per unit of electricity generated over a power plant's lifetime, accounting for all capital, operating, and fuel costs. Using LCOE for comparison, renewable energy sources are now among the cheapest new-build electricity generation methods, with unsubsidized utility-scale solar

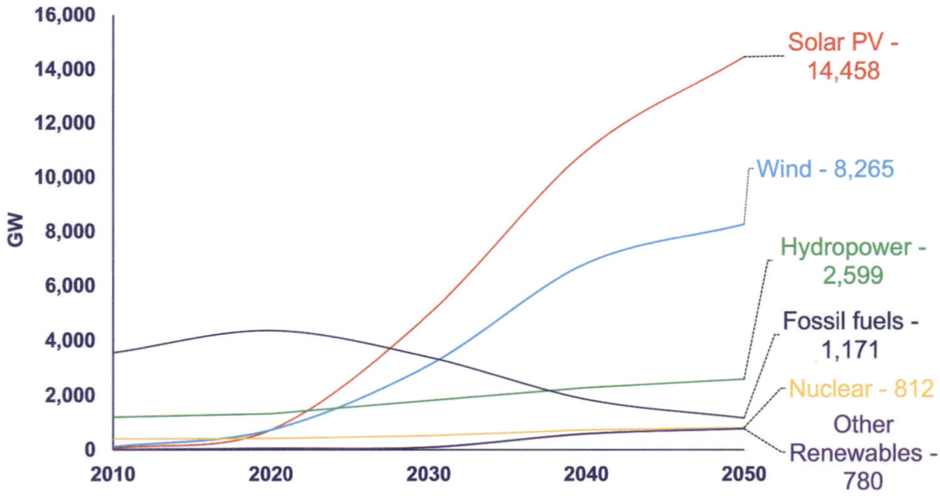


**Fig. 1.1** The evolving trend of world energy resources for electricity, with an outlook towards 2050 (per IEA)

typically costing between \$38–78/MWh and onshore wind between \$37–86/MWh [7]. Renewables other than solar and wind are more expensive—geothermal energy has a cost of approximately \$66–109/MWh, and large hydropower typically falls between 40–110/MWh [9], and the deployment of these power plants is selective on-site conditions, also limiting their scalability [8]. On the other hand, fossil fuel plants are generally more expensive than renewables. Natural gas peaking plants, cost roughly \$149–251/MWh, and coal plants cost around \$71-173/MWh [8].

A key economic advantage of renewables is that once a solar or wind plant is built, it requires no ongoing fuel input, meaning operating costs are minimal and largely fixed over time. Conversely, fossil fuel generators rely on continuous fuel purchases—whether coal, gas, or oil—adding a significant variable cost component subject to market price swings. Nuclear energy, while it does not generate pollution, remains significantly more expensive at \$141–280/MWh due to high capital costs, long construction times, and stringent regulatory requirements [8]. The above reasons render solar and wind being recognised to be two leading sources of energy in the future, predicted to reach over 14,000 GW and 8000 GW, respectively.

However, one of the emerging challenges in scaling up solar and wind energy systems is the competition for space. Large-scale solar farms and wind arrays require vast tracts of land: a typical 1 GW solar installation may occupy more than 15–20 $km^2$, while onshore wind farms, due to more space required between individual turbines and for surrounding areas, can require around 100 square kilometers. This growing demand for land often conflicts with agriculture, urban expansion, conservation efforts, and indigenous rights, leading to public resistance and project delays. The land-use conflict becomes particularly acute in densely populated regions or areas with high biodiversity value, where renewable expansion is environmentally or socially constrained.

In response to these limitations, the energy sector is increasingly turning to the ocean as a new frontier for sustainable energy development. The ocean covers over 70% of the Earth's surface, offering vast, untapped space for renewable deployment. Offshore wind farms, floating solar platforms can operate far from human settlements, minimizing land conflicts while harnessing clean energy. In addition, the constant movement of ocean flow also provides additional sources of energy, having inspired the development of wave and tidal energy converters. Ocean renewable energy not only relieves pressure on land use but also opens up possibilities for collaboration with the offshore networks to enhance production and transportation, creating multi-use energy hubs.

Ocean-based energy system is not just a vision. In fact, there have been over 85 GW offshore wind farms already installed [10], followed by significant progress of ocean-based solar, wave, and tidal systems. As offshore technologies mature and become more cost-effective, they are poised to play a central role in the ongoing energy transition.

## 1.2 Challenges of Ocean Renewable Energies

As the world pivots toward a low-carbon energy future, the oceans are increasingly seen as the next frontier for large-scale renewable energy deployment. Offshore environments offer abundant natural resources—strong winds, relentless waves, steady tides, and vast open spaces for solar panels—but also introduce formidable challenges that remain unresolved. Unlike their onshore counterparts, offshore renewable systems must operate in dynamic, corrosive, and often remote marine environments. Designs that can sustain in such conditions are still evolving and far from mature.

One major challenge lies in mechanical survivability in harsh marine conditions. Offshore wind turbines, wave/tidal energy converters, and floating solar platforms must resist the combined (a) forces such as strong winds, ocean currents, cyclic and extreme waves, (b) saltwater corrosion, (c) biofouling from marine organisms such as algae, barnacles, and mussels that increases drag, adds weight, and impairs device functionality, and (d) seabed scouring where sediments are removed around the foundations of energy systems fixed on the seabed. These effects can lead to dysfunctional, fatigue, material degradation, fracture, and even catastrophic structural failure, placing demands on structural integrity and durability. A majority of contemporary prototypes of offshore wind, wave, tidal and solar devices are still in the early commercialization phase. Also, general infrastructure such as dynamic cables and mooring that connect floating platforms to the grid and seabed still suffer from frequent fatigue-induced failures, causing significant costs in maintenance.

Predictive methods for environmental and operational loads in ocean-based renewable energy systems remain insufficiently accurate, particularly under extreme or site-specific conditions. Contemporary models often struggle to account for the complex interactions between wind, waves, currents, and structural dynamics, especially for floating platforms that undertake sea-induced motions. As a result, designers tend to adopt conservative safety margins to compensate for uncertainty, leading to over-engineered and cost-inefficient systems. This conservatism increases structural mass, material use, and installation complexity, without proportionate gains in reliability. Moreover, the lack of accurate load prediction impairs both operational optimisation and asset management. Lacking of real-time monitoring of comprehensive loading, for instance, reduces the effectiveness of condition-based maintenance strategies. Servicing teams are often not sent out offshore at optimal times. Similarly, energy yield predictions become less reliable, undermining grid integration planning and economic modelling.

Compared to onshore systems, cost remains a major barrier for offshore renewable projects. For example, the levelized cost of electricity (LCOE) for offshore wind is around 70–157/MWh, compared to $37–86/MWh for onshore wind [8]. For emerging technologies like offshore wave, tidal and solar, the LCOE is often still not economically viable due to limited deployment scale and technological immaturity.

Insallation expenditures for offshore energy systems are significantly higher than for land-based, largely due to complex logistics, specialized equipment, and the high-risk nature of offshore construction. Installation requires specialized vessels such as jack-up barges or dynamically positioned ships. These operations are further constrained by narrow weather windows; rough seas and high winds limit construction and maintenance activities to just a few months per year in many regions. Once operational, offshore systems face significantly higher operation and maintenance (O&M) expenses compared to onshore equivalents due to difficult access, the need for specialist labor, and heavy reliance on support vessels. Even routine inspections, including blade condition assessments, cable integrity checks, or corrosion monitoring, typically require drones, remotely operated vehicles (ROVs), or human divers under challenging and hazardous conditions, adding further complexity and cost. Without innovation in condition monitoring systems, predictive maintenance, and modular designs, many offshore solutions remain economically unfavourable.

Digitalization, now a cornerstone of optimization in land-based renewable energy systems, remains significantly underdeveloped in offshore environments. While onshore wind farms benefit from high-bandwidth telemetry, offshore platforms face a range of digital integration challenges. Connectivity is a major limitation, with satellite communication still the default in many remote offshore areas, resulting in low data throughput and high latency. Underwater environments limit wireless data transmission. Additionally, sensors used for structural health monitoring and environmental data collection often degrade rapidly due to harsh marine conditions. These issues contribute to frequent data gaps, which undermine the effectiveness of predictive maintenance strategies and hinder system optimization. As a result, designers often resort to conservative safety margins that drive up both cost and structural weight. Although there is growing interest in deploying AI-driven digital twins, fault prediction algorithms, and autonomous underwater monitoring systems, these technologies remain in the early stages of field validation for offshore applications.

A central challenge facing the sector is that many offshore renewable energy systems are still adapted from land-based designs rather than being conceived specifically for the unique conditions of marine environments. This legacy approach limits performance and scalability. What is needed is a systemic rethinking that embraces ocean-specific innovation. This includes the development of modular and scalable architectures that allow for rapid deployment, replacement of individual components, and relocation of offshore systems. The use of advanced materials, such as embedded sensors' encapsulation, self-healing composites, nano-structured corrosion-resistant alloys, and marine-grade coatings, can significantly improve durability by resisting biofouling and saltwater damage. Autonomous O&M systems, including drones and autonomous underwater vehicles (AUVs), can help reduce human intervention and increase uptime by enabling continuous, low-cost inspection and repair. Hybrid marine platforms, which integrate multiple generation technologies such as wind, wave, tidal, and solar power on a single floating structure, offer a path toward higher energy density and system resilience. To manage the complexity of these future systems, next-generation energy management platforms are also needed. These must be capable of dynamically balancing power output across hybrid assets, integrating with grid-scale energy storage, and optimising operations for both energy yield and ecological sustainability.

## 1.3 Scope and Structure of This Book

The formidable technical and economic challenges outlined in the previous section underscore a pressing reality: while offshore renewable energy holds immense promise, its path to large-scale viability is still obstructed by a range of unresolved issues. Mechanical survivability in marine conditions, underdeveloped predictive models, costly deployment and maintenance procedures all point to a need not just for incremental improvement, but for systemic innovation.

This book was conceived with that gap in mind. It responds directly to the challenges discussed in Sect. 1.2 by exploring both contemporary solutions and forward-looking strategies. The aim is not merely to document the current state of offshore renewable energy systems, but to critically evaluate where the field stands today, where it needs to go, and how it might get there. The book achieves this by presenting a holistic, interdisciplinary view that combines engineering realities with environmental insights and digital systems thinking.

By analysing existing barriers—such as the lack of accurate load prediction, high O&M costs, limited digital integration, and insufficiently marine-specific design philosophies—we set the stage for examining the technological advances, system-level innovations, and collaborative models that may overcome them. Rather than treating offshore renewables as an extension of onshore systems, this book emphasizes the unique opportunities and demands of ocean-based energy, encouraging a new design mindset: one that is modular, resilient, digitally intelligent, and economically feasible.

This book is structured to guide the reader from foundational knowledge to emerging opportunities. Chapters 2 through 5 each focus on a major offshore renewable technology, dissecting the principles, components, design strategies, deployment challenges, and promising innovations associated with each. Chapter 6 presents potentials of how the technologies in each previous chapter can complement each other and connect to the wider ocean ecosystem.

Chapter 2 focuses on offshore wind turbines, the most mature and widely deployed offshore renewable technology. It introduces the operating principles of offshore wind turbines, explores various designs and subsystems, and offers a global overview of current deployment. The chapter concludes by identifying key technical and operational challenges, such as structural design, wake interactions, and service at sea.

Chapter 3 examines wave energy converters, which harness the oscillatory motion of waves to generate electricity. After introducing the physical principles behind wave energy, it provides a detailed classification and assessment of wave energy devices, key design considerations, and the current state of technological development across various actors in the sector.

Chapter 4 turns to tidal energy systems, which extract energy from predictable tidal flows. The chapter compares turbine and other designs, analyses site suitability in different geographic and hydrodynamic contexts, and addresses the environmental and engineering constraints that still limit scalability.

Chapter 5 is dedicated to floating ocean solar systems. It introduces the principles and configurations of marine-based photovoltaic systems and explores the current and future trends in design and integration. Particular attention is given to the modularity and survivability of these systems in open-sea conditions.

In Chap. 6, this book envisions how different types of offshore energy technologies can be integrated into cohesive multi-use platforms, and their incorporation with a wider offshore ecosystem such as maritime networks, aquaculture, and fuel production. It also addresses cross-cutting themes like digitalisation, predictive maintenance, and intelligent systems.

Finally, Chap. 7 serves as an epilogue and reflection. It discusses the role of offshore renewables in a rapidly evolving global energy landscape and calls for a new generation of thinkers and practitioners who can bring creativity, collaboration, and critical thinking to this field.

This book is intended for engineers, researchers, students, and decision-makers alike. Whether your interest lies in technical detail, system-level integration, or future innovations, this book is expected to not only inform but also inspire. By illuminating both the status quo and the emerging possibilities, we invite readers to engage deeply with the ocean energy sector—not just as it is, but as it could be.

## References

1. Ritchie H. & Rosado P. "Energy Mix." Our World in Data. Global Change Data Lab (2020).
2. Forster P. M., Smith C., Walsh T., Lamb W.F., Lamboll R., Cassou C., Hauser M., et al. Indicators of Global Climate Change 2024: annual update of key indicators of the state of the climate system and human influence. Earth Syst. Sci. Data 17, 2641–2680 (2025).
3. European Environment Agency. Global and European sea level rise: Indicator assessment. Published 14 Jan 2025.
4. World Meteorological Organization (WMO). Atlas of Mortality and Economic Losses from Weather, Climate and Waterrelated Hazards (1970–2021). May 2023.
5. Lebling K., Gangotra A., Hausker K. & Byrum Z. 7 Things to Know About Carbon Capture, Utilization and Sequestration. World Resources Institute, 16 May 2025.
6. IEA. World Energy Outlook 2022. International Energy Agency, Paris (2022).
7. Kennedy R. Solar cost of electricity beats lowest-cost fossil fuel – even without tax credits. pv-magazine USA, 1 July 2025.
8. Lazard Inc. Levelized Cost of Energy+ (LCOE+) — June 2025. Version 18.0. New York: Lazard, June 2025.

9. Edghill L. How much does hydroelectric power cost per unit? Green Business Barbados (2025).
10. RenewableUK. Global Offshore Wind Pipeline — June 2025. EnergyPulse report, 16 June 2025.

# 2 Offshore Wind Turbines

Over the past two decades, onshore wind turbines have grown significantly in both height and generating capacity. Early-generation onshore wind towers stood approximately 60 m tall plus rotor (blades and a central hub) diameters of about 80 m, while modern wind towers now routinely exceed 100 m in total height and are equipped with rotor diameters of more than 120 m. This scale is driven by the pursuit of greater efficiency and lower energy costs, as larger rotors capture more wind and reduce the LCOE.

However, the increasing size of onshore turbines has brought growing tension with the human and ecological environment. In rural or semi-rural regions, residents often raise concerns about visual intrusion, noise, and shadow flicker caused by the rotating blades. The presence of tall, moving structures in previously open landscapes can alter the visual character of a region, leading to opposition rooted in aesthetic, cultural, or property value considerations. While modern turbines are quieter than earlier models due to improved aerodynamic designs, low-frequency noise and amplitude modulation effects can still cause annoyance or sleep disruption, particularly in communities located within 1–2 km of wind farms.

Bird and bat mortality is another major concern associated with onshore wind turbines. Collisions with fast-moving blades pose significant threats to certain bird species, especially raptors, and migratory birds flying at rotor height. The cumulative impact of large onshore wind farms on biodiversity, particularly when sited along flyways or near protected habitats, has prompted ecological studies and triggered public and regulatory scrutiny. Moreover, constructing and maintaining land-based wind farms often requires access roads, site clearance, and disturbance to terrestrial ecosystems.

These challenges have begun to limit the availability of suitable land for wind development, especially in densely populated or environmentally sensitive regions. As a result, the energy sector is increasingly turning its attention offshore, where wind resources are stronger, space is abundant, wildlife density is lower, and human presence is minimal.



L. Huang, *Ocean Renewable Energy Systems*, Synthesis Lectures on Renewable Energy Technologies, https://doi.org/10.1007/978-3-032-10576-9_2

Offshore wind turbines can be built farther from residential areas, reducing visual and acoustic impacts, and can be placed in zones that avoid critical bird migration corridors or ecologically sensitive coastlines. By relocating wind infrastructure to the sea, developers can reduce land-use conflicts and environmental trade-offs, while also taking advantage of higher wind speeds and steadier flows. Consequently, offshore wind turbines can be larger than onshore and yield more power.

Offshore wind, therefore, is not merely a technical alternative to onshore systems—it represents a strategic spatial shift in how renewable energy is developed. The oceans offer a new frontier for clean power, potentially supporting much larger turbines and more expansive arrays without the same degree of social and ecological resistance faced on land. Yet, the offshore environment also brings its own unique challenges, from structural survivability to monitoring and maintenance, which are the focus of the sections that follow.

## 2.1 Operating Principles

Offshore wind turbines convert the kinetic energy of wind over the sea into electrical energy. The fundamental principle is similar to that of onshore turbines: wind turns the rotor blades, which spin a shaft connected to a generator that produces electricity. However, the offshore environment introduces distinct characteristics and advantages, such as stronger and more consistent wind speeds, leading to higher energy yield. Contemporary onshore wind turbines generally have rated power ranging from 3 to 8 MW, while offshore ones can achieve 10–20 MW. The overall size of wind turbines, especially offshore, is still getting larger, and this trend is expected to continue, though it is beginning to slow down as the industry nears practical, logistical, and economic limits.

There are two general types of offshore wind turbines, classified based on how they are connected to the seabed: fixed-bottom and floating. The choice between these configurations primarily depends on the water depth and seabed conditions at the project site. Fixed-bottom offshore wind turbines are the most established and widely deployed. They are typically installed in water depths of up to 60 m, where conditions allow for stable foundation construction. Fixed-bottom turbines benefit from well-developed installation techniques and no degree of freedom to reduce uncertainties. However, their deployment is constrained to continental shelf regions, which limits access to higher wind resource areas further offshore. These turbines are fixed directly to the seabed using foundation types such as monopiles, jackets, suction buckets, or gravity-based structures. Fixed-bottom offshore wind utilises turbines in the 8–16 MW range with tower heights around 130 m and blade lengths on the order of 90-120 m. Floating offshore wind enables deployment in much deeper waters, using floating platforms anchored to seabed via mooring, which enables the structural sizes to be much bigger than fixed-bottom ones. Floating wind capacities can be over 20 MW, reaching rotor diameters of over 300 m [1].

As shown in Fig. 2.1, an offshore wind turbine system includes (a) Rotor and Blades: capture kinetic energy from the wind; (b) nacelle: Houses the gearbox, generator, controller, and braking systems; (c) Tower: Supports the nacelle and rotor; (d) Foundation/floating structures: supporting the superstructure above water—fixed-bottom types include monopile, jacket, suction caisson, gravity-based foundations, and floating types of floaters such as spar-buoy, semi-submersible, tension-leg along with mooring and anchoring—this is the most distinguished difference between fixed-bottom and floating wind turbines; (e) Electrical Infrastructure: includes internal and external cables.

The main types of foundation for fixed-bottom offshore wind turbines are illustrated in Fig. 2.2. Globally, monopile foundations dominate the fixed-bottom offshore wind market, accounting for approximately 80% of installations due to their simplicity, cost-effectiveness, and suitability for shallow waters up to 50 m deep. Jacket structures represent around 10–15%, typically used in deeper waters or where seabed conditions demand more robust support. Gravity-based foundations make up about 5%, although their use is limited by logistical and seabed preparation requirements. Suction bucket foundations hold a minimal share, but are gaining attention for their low-noise installation, ease of decommissioning, and compatibility with soft seabeds. The distribution reflects both technological maturity and geographic deployment patterns.

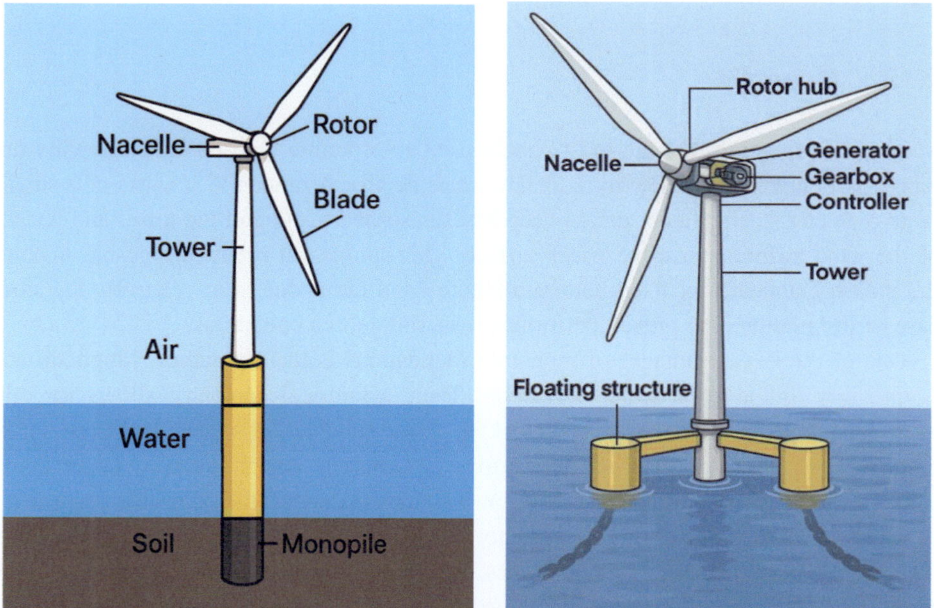


**Fig. 2.1** Illustration of a fixed offshore wind turbine (left) and a floating wind turbine (right)

**Fig. 2.2** Fixed-bottom offshore wind turbines with different types of foundations

### 2.1.1 Monopile

Monopile foundations are the most widely used substructure type for offshore wind turbines, particularly in shallow to moderate water depths. A monopile consists of a single, large-diameter steel cylinder driven deep into the seabed to support the tower and nacelle of the wind turbine above the water surface. This simple yet robust design has become the industry standard for fixed-bottom offshore wind farms due to its relatively low cost, ease of installation, and proven performance in various sea conditions.

One of the key advantages of monopile foundations is their structural simplicity and adaptability to industrial-scale fabrication. Their cylindrical geometry allows for efficient load transfer into the seabed, making them suitable for water depths of up to approximately 30–40 m. Recent innovations in monopile design, such as larger diameters (up to 10 m) and advanced steel grades, have extended their feasible application to even deeper waters and higher-capacity turbines. Compared to more complex foundation types like jackets or gravity bases, monopiles offer reduced manufacturing and installation time, especially when combined with specialized vessels such as jack-up rigs and pile-driving hammers. Their standardization and scalability make them attractive for large-scale deployment.

However, monopiles also have limitations. Their structural efficiency decreases in very deep waters or soft seabed conditions, where lateral and vertical stability become more difficult to ensure. In such environments, alternative foundations like jackets, suction buckets, or floating platforms may be more appropriate. Additionally, pile driving during installation can generate significant underwater noise, which poses risks to marine mammals and requires mitigation measures such as bubble curtains or acoustic dampening. The decommissioning and recyclability of monopiles also present long-term environmental and logistical challenges that are still being addressed.

Despite these drawbacks, monopiles continue to dominate the global offshore wind market. In Europe, the world's leading region for offshore wind development, monopiles account for more than 70% of installed turbine foundations. For example, the Hornsea 2 offshore wind farm in the UK uses monopile foundations to support its 165 turbines, delivering a total capacity of approximately 1.4 GW. Similarly, the Borssele 1 & 2 wind farms off the Dutch coast utilize 94 monopile-mounted turbines with a combined output of 752 MW.

Going forward, the future potential of monopile foundations remains strong, particularly as turbine ratings continue to increase and cost reduction remains a key driver of offshore wind competitiveness. Advanced installation techniques, including vibratory pile driving and low-noise methods, are under development to mitigate environmental impacts. In parallel, hybrid solutions such as monopile-suction caisson combinations are being tested to enhance geotechnical performance in challenging seabeds. With global ambitions to expand offshore wind capacity, monopiles are expected to play a major role in the next generation of fixed-bottom offshore wind farms, at least in waters up to 50 m deep.

### 2.1.2 Jacket

Jacket foundations are a type of fixed-bottom substructure used to support offshore wind turbines, particularly in deeper waters where monopile foundations become structurally inefficient. A jacket foundation consists of a lattice-like steel frame with typically three or four legs connected by diagonal braces, resembling an open truss structure. These legs are anchored to the seabed using piles, providing a wide base that offers excellent stability and load distribution. Originally developed for the oil and gas industry, jacket foundations have been adapted for offshore wind as turbines grow larger and projects move into deeper and more energetic waters.

One of the primary advantages of jacket foundations is their suitability for intermediate to deep water depths, typically ranging from 30 to 60 m, and potentially up to 80 m with appropriate design. Their open framework structure allows for reduced hydrodynamic loading, as water can flow through the lattice, rather than pushing against a solid wall; as in the case of monopiles. This makes jackets particularly advantageous in areas with strong currents or wave action. Additionally, the distributed load-bearing design offers better geotechnical performance in complex seabed conditions, such as rocky or uneven terrain where monopiles may be difficult to install.

Despite these strengths, jackets are more complex to design, and costly to manufacture and install compared to monopiles. The lattice structure requires precise fabrication and high-quality welding, and the installation process is more intricate due to the need for multiple pin piles and precise alignment. As a result, jackets are generally reserved for deeper waters where monopiles are no longer cost-effective. Maintenance and inspection of jacket foundations can also be more challenging due to the increased number of structural elements and potential points of corrosion or fatigue. Furthermore, the greater amount of steel used in jackets compared to monopiles can result in a higher carbon footprint unless offset by long-term performance and recyclability.

Although jackets are less common than monopiles globally, they have played a key role in many high-profile offshore wind projects. Jackets are preferred beyond ~45 m depth or in high-seismic-risk regions. In the UK, the Beatrice Offshore Wind Farm in the Moray Firth uses 84 Siemens Gamesa 7 MW turbines supported by jacket foundations in water depths of up to 56 m, delivering a total capacity of 588 MW. Another example is the Wikinger wind farm in the German Baltic Sea, which utilizes 70 Adwen 5 MW turbines on jacket structures, contributing 350 MW to the grid. In Asia, jacket foundations are gaining traction in markets like Taiwan and Japan, where seismic activity, typhoons, and complex bathymetry necessitate highly resilient structures.

Looking to the future, jacket structures allows for optimization and potential cost reductions through economies of scale and advanced fabrication methods. Ongoing research into lightweight materials, corrosion-resistant coatings, and automated welding technologies could further improve their economic competitiveness. While floating foundations are poised to dominate in very deep waters beyond 60–80 m, jackets will likely continue to serve as the preferred fixed-bottom solution in transitional zones—bridging the gap between shallow monopile-suitable sites and areas requiring floating technologies.

### 2.1.3 Suction Bucket

Suction bucket foundations, also known as suction caissons, are an innovative form of fixed-bottom substructure used to support offshore wind turbines. These foundations consist of large, inverted steel cylinders (buckets) that are installed by lowering them onto the seabed and creating a pressure differential through suction. As water is pumped out from inside the bucket, the resulting pressure drop forces the structure into the seabed, securing it in place without the need for pile driving. Suction buckets can be used alone or as part of a jacket structure, depending on the site conditions and turbine size.

The main advantages of suction bucket foundations lie in their quiet, fast, and reversible installation process, which avoids the high noise levels and seabed vibrations associated with traditional pile driving. This makes them particularly attractive for areas with strict environmental regulations or sensitive marine habitats. The installation is relatively quick and requires fewer heavy-lift operations, contributing to lower vessel time and reduced installation costs. Suction buckets also allow for easy decommissioning or relocation, as the same suction mechanism can be reversed to lift the structure out of the seabed without leaving behind embedded materials.

From an engineering standpoint, suction bucket foundations offer excellent stability against lateral loads, making them suitable for soft or muddy seabeds where other foundation types might struggle. They are also promising for transitional water depths between those suited to monopiles and floating structures. However, their use is still relatively limited due to site-specific constraints, such as the need for appropriate soil permeability and cohesion. In highly sandy or rocky seabeds, achieving the required suction seal and penetration depth can be problematic. Additionally, the technology is less mature than monopile or jacket foundations, and requires advanced geotechnical analysis to ensure safe long-term operation.

Several pilot and commercial-scale projects have already demonstrated the viability of suction bucket technology. A notable example is the Borkum Riffgrund 1 project in the German North Sea, where three 4 MW Siemens turbines were installed on bucket foundations as part of a demonstration in 2014. Similarly, at the Aberdeen Bay in Scotland, 11 Vestas 8.4 MW turbines were installed on suction bucket jackets in 2018, marking the first commercial-scale use of the technology in the UK. These foundations were chosen to minimize environmental impact and to streamline the construction timeline in a region known for complex permitting conditions.

Looking forward, suction bucket foundations have significant potential in the offshore wind sector, especially as developers seek to reduce the environmental footprint of projects and increase deployment efficiency. As turbine sizes continue to grow, the ability to install large structures with minimal seabed disturbance becomes increasingly valuable. Additionally, improvements in computational soil modeling could enhance design accuracy and broaden the range of seabed conditions suitable for suction installation.

### 2.1.4 Gravity-Based

Gravity-based foundations (GBFs) are one of the earliest types of substructures used for offshore wind turbines. As the name implies, these foundations rely on their own weight—typically made from concrete, sometimes with added ballast—to remain stable on the seabed without the need for piling or anchoring. The base of the structure is wide and heavy, designed to resist the overturning moments and lateral forces generated by waves, wind, and turbine operation. Gravity-based designs are usually prefabricated onshore, towed to site, and then placed onto a prepared seabed, often involving levelling and reinforcement with gravel or rock.

One of the main advantages of gravity-based foundations is that they require no pile driving, eliminating underwater noise pollution and the need for deep penetration into the seabed. This makes them particularly attractive for regions with strict environmental regulations or sensitive marine ecosystems. Additionally, GBFs are highly durable and have long lifespans due to the use of reinforced concrete, which is resistant to corrosion and fatigue. Their onshore fabrication allows for better quality control and reduces the need for complex offshore assembly. GBFs are also well-suited to nearshore and ice-prone environments, where their stability and mass help resist ice loading and movement.

However, gravity-based foundations also come with notable drawbacks. Their massive size and weight (often requiring thousands of tonnes of concrete per unit) demand significant seabed preparation and strong logistical support for transportation and installation. The need for stable, relatively flat seabeds limits the range of suitable deployment sites. In deeper waters, their size and construction complexity scale up rapidly, making them economically less competitive than lighter alternatives like monopiles or jackets. Moreover, the use of large volumes of concrete raises environmental concerns related to embodied carbon emissions and construction impacts.

Despite these limitations, gravity-based foundations have been successfully deployed in several commercial projects. For example, Middelgrunden offshore wind farm near Copenhagen, Denmark, which began operation in 2001 with 20 Bonus 2 MW turbines supported by gravity bases in shallow water (only 3–8 meters). Another notable project is the Thornton Bank Pilot Phase in Belgium, which used GBFs for six 5 MW turbines in 2009, before later phases transitioned to jacket structures. These projects demonstrate the suitability of GBFs in nearshore, shallow, or geotechnically stable areas.

Looking ahead, gravity-based foundations may regain attention as part of the broader drive to diversify foundation solutions and reduce environmental disturbance. Innovations such as modular GBFs, use of low-carbon concrete, and hybrid designs combining concrete with steel components are under exploration to improve feasibility and sustainability. GBFs may also play a role in hybrid marine structures, such as combining wind turbines with breakwaters or port extensions. In countries with limited heavy offshore installation infrastructure, onshore prefabrication and float-out methods may offer strategic advantages.

### 2.1.5 Spar-Buoy

Floating offshore wind turbines (FOWTs), by contrast to fixed offshore wind turbines, are designed for deep waters (typically beyond 60 m), where fixed foundations become economically or technically unfeasible. Instead of anchoring directly to the seabed, these turbines are mounted on buoyant platforms that are moored in place using tensioned cables or chains. As shown in Fig. 2.3, several types of floating platforms exist, including spar-buoy, semi-submersible, and tension-leg designs, each with different

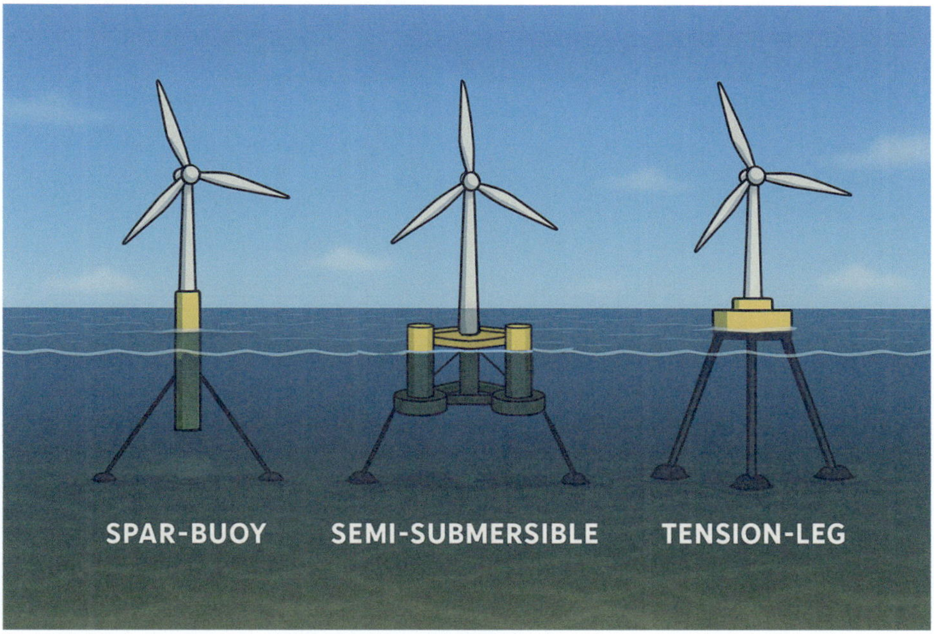


**Fig. 2.3** Illustration of typical designs of floating offshore wind turbines

suitabilities. Floating turbines unlock vast areas of the ocean previously inaccessible, particularly deep seas with high wind speeds. Although still emerging as a commercial technology, floating wind has rapidly gained traction due to its scalability and potential to avoid visual and environmental impacts associated with nearshore installations. FOWTs are subject to continuous motion from ocean waves, which influence their stability, structural integrity, power generation performance, and control system design. Unlike fixed-bottom turbines, floating platforms are not rigidly attached to the seabed, so they respond dynamically to environmental forces through six degrees of freedom: surge, sway, heave, roll, pitch, and yaw, as illustrated in Fig. 2.4. Pitch and heave are the most critical for turbine performance. Excessive pitch motion can cause rotor misalignment, reducing aerodynamic efficiency and increasing control complexity. Heave motion impacts drive-train loading and clearance between blade tips and water surface (especially in high sea states).

Spar-buoy floating offshore wind turbines are one of the most mature and proven types of floating wind foundation technologies, designed for deep-water sites where fixed-bottom structures are not economically or technically feasible. A spar-buoy consists of a tall, slender cylindrical hull that extends deep below the water surface and is ballasted at the bottom to provide stability. The turbine tower is mounted atop the floating spar, and the entire structure is kept in position using a mooring system anchored to the seabed.

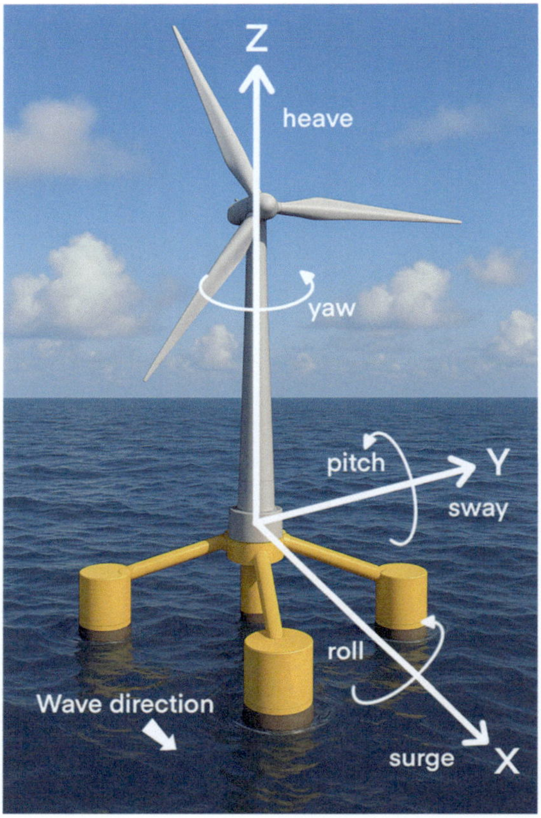


**Fig. 2.4** Illustration of wave-induced motions of a floating wind turbine

The spar's deep draft ensures a low center of gravity and strong resistance to pitch and roll motions caused by waves and wind.

One of the key advantages of spar-buoy designs is their exceptional stability in harsh ocean conditions. Because much of the buoy's volume lies beneath the surface, it experiences minimal wave-induced motion compared to other floating platforms like semi-submersibles. This improves turbine performance and reduces fatigue on structural components. Spar-buoys also offer a relatively simple geometric design, which can be advantageous for manufacturing, and their hydrodynamic behavior is well understood, thanks to extensive deployment in the offshore oil and gas industry. Additionally, spar-buoys have minimal surface footprint, which reduces visual impact and can ease permitting challenges in some coastal zones.

However, spar-buoy platforms have limitations. Their deep draft (around 60–120 m) can increase material costs as well as difficulty in manufacture and deployment. The ballast required to maintain stability adds significant weight, and the tall hull structures can be expensive and complex to fabricate and transport. Moreover, spar-buoys are best suited to deep-water locations typically over 100 m, making them less viable for transitional zones or nearshore installations.

The spar-buoy concept has been successfully demonstrated at utility scale, most notably through Equinor's Hywind Scotland project, which became the world's first commercial floating wind farm when it was commissioned in 2017. Developed by Equinor, Hywind Scotland features five Siemens Gamesa 6 MW turbines, each mounted on a spar-buoy structure, totaling 30 MW of installed capacity. Located 25 km off the coast of Peterhead in water depths of around 95–120 m, the project has achieved capacity factors exceeding 50% (actual energy output divided by maximum possible output), outperforming many fixed-bottom wind farms. Building on this success, Equinor developed Hywind Tampen, a larger 88 MW floating wind farm in the North Sea that that supplies power to offshore oil platforms. The project uses 11 turbines rated at 8.6 MW each based on the spar-buoy design in 260–300 m deep waters, and was commissioned in 2023.

In the future, spar-buoy foundations are poised to play a central role in the expansion of floating wind energy, especially in deep-water regions with consistent wind resources. As turbine sizes increase, the spar's stable platform is particularly well-suited. Innovations in modular construction, materials, and ballast optimization could reduce fabrication and transport costs. In addition, the development of specialized deep-water ports for floating wind assembly could unlock further deployment at scale. However, spar-buoys will continue to face competition from alternative floating technologies like semi-submersibles and tension-leg platforms.

### 2.1.6 Semi-submersible

Semi-submersible floating offshore wind turbines represent one of the most versatile and widely adopted floating foundation concepts for deploying wind turbines in deep waters. A semi-submersible platform consists of multiple interconnected floating columns (typically three or four) that provide buoyancy and stability through distributed displacement and ballast. These columns are partially submerged and linked by a structural framework, upon which the wind turbine tower is mounted. The system is held in position using a spread mooring system, typically involving catenary or taut lines anchored to the seabed.

The primary advantage of semi-submersible platforms lies in their stability in a wide range of water depths. Unlike spar-buoy platforms, semi-submersibles have a shallower draft, allowing them to be assembled and floated out from shallower ports, which simplifies logistics and reduces port infrastructure requirements. Their horizontal footprint allows for turbine integration at quayside, meaning the turbine can be mounted onto the platform onshore and then towed into place fully assembled. This approach improves quality control and reduces expensive offshore work.

However, semi-submersibles also face specific challenges. Their larger surface area above the waterline exposes them to more wave action, leading to more complex dynamic behavior compared to spar-type structures. As a result, more sophisticated control systems and ballast management may be required to ensure stable turbine operation. The structural complexity and larger platform area also translate to higher material use and construction costs. Moreover, their mooring and anchoring systems must be carefully designed according to the platform design, which can be more expensive and environmentally impactful in some seabed conditions.

Several high-profile projects have demonstrated the viability of semi-submersible floating wind turbines. One of the earliest commercial-scale examples is the WindFloat Atlantic project off the coast of Portugal. Commissioned in 2020 and developed by Principle Power and a consortium of partners, the project consists of three Vestas V164-8.4 MW turbines mounted on semi-submersible platforms, totaling 25.2 MW. The floating platforms were assembled in port and towed more than 20 km offshore to water depths around 100 m. The WindFloat design has since gained attention for its adaptability and relatively simple deployment. Another notable project is Kincardine Offshore Wind Farm in Scotland, developed by Cobra Group. Operational since 2021, it includes five 9.5 MW turbines (GE Haliade), mounted on semi-submersible structures.

In the years to come, semi-submersible platforms are expected to play a central role in scaling up floating offshore wind due to their relative adaptability, proven performance, and compatibility with existing shipyard infrastructure. As turbine sizes increase and floating projects move into deeper waters and larger arrays, semi-submersibles offer a scalable and manufacturable solution. Advances in modular platform construction, mooring designs, and automated installation methods may further reduce lifecycle costs and improve competitiveness.

### 2.1.7 Tension-Leg

Tension-leg platforms (TLPs) are a distinctive type of floating foundation used in offshore wind energy, especially suited for deep-water deployment. Unlike spar-buoy or semi-submersible platforms that rely primarily on buoyancy and mooring systems for stability, TLPs achieve their stability through vertical tensioned mooring lines, also known as tendons, which are anchored firmly to the seabed. The floating platform itself is buoyant and sits above the water surface, but the taut tendons keep it in a fixed vertical position, minimizing pitch and heave motions. This unique mooring method allows for a lightweight, compact platform design with excellent stability.

The main advantage of TLPs lies in their exceptional motion stability, particularly in the vertical direction. This makes them especially suitable for supporting large offshore wind turbines where minimizing nacelle movement is critical for efficient operation and component longevity. Additionally, because TLPs are anchored with vertical mooring lines, they require less horizontal footprint than catenary moored systems, which is beneficial in areas with crowded seabeds or existing subsea infrastructure. Moreover, TLPs can be designed with shallow drafts, allowing for easier integration and assembly in a wider range of ports.

Despite these benefits, TLPs present several challenges. The installation process is complex, as it requires precise tendon tensioning and anchoring, often needing specialized vessels and careful sequencing to achieve neutral buoyancy and platform stability. This can raise upfront costs and limit deployment to regions with adequate infrastructure and offshore installation capability. Additionally, seabed anchoring requirements are more stringent for TLPs, requiring high-strength foundations to accommodate vertical loads. This makes them less suitable for regions with soft or unconsolidated seabeds. Their dynamic response to extreme conditions must also be carefully modeled, particularly in regions with strong currents or significant swell.

Several demonstration projects have explored the use of TLPs for floating wind. Keuka Energy tested a vertical-axis floating turbine prototype in the U.S., which used tension legs as part of its stabilization. The GICON-SOF (Floating Offshore Foundation) in Germany used a hybrid TLP design and has undergone scale-model testing. While no full-scale commercial TLP-based wind farm has yet been constructed, several developers, including SBM Offshore, Technip Energies, and Glosten (PelaStar), have proposed TLP-based systems for future deployment with 10–15 MW turbines.

Looking to the future, tension-leg floating platforms hold significant promise for the next generation of offshore wind farms, especially in deep-water regions with high wind potential and limited port depth for spar deployment. Their compact footprint and minimal motion make them well-suited for use with large-scale wind turbines, which are sensitive to platform movement. Advances in tendon material, automated installation processes, and hybrid platform designs could help reduce costs and broaden applicability.

## 2.2 Offshore Wind Farm

### 2.2.1 Typical Array Layouts

Offshore wind turbines are usually arranged in geometric patterns (arrays) to optimise energy capture and ease operations, known as offshore wind farms. Many projects use a grid layout consisting of turbines in rows and columns. Within this grid, two common configurations are an aligned layout and a staggered layout. Figure 2.5 provides examples of some renowned offshore wind farms around the world.

Aligned layout—turbines in each row sit directly behind those in the front row (along the prevailing wind direction). This regular spacing is simple for planning and cabling, but it means each downstream turbine operates in the full wake of an upstream turbine. Aligned rows can therefore suffer greater energy losses, as second-row turbines face a lower wind speed and generate less power. This layout is sometimes chosen when site geometry or uniform wind direction makes straight rows practical.

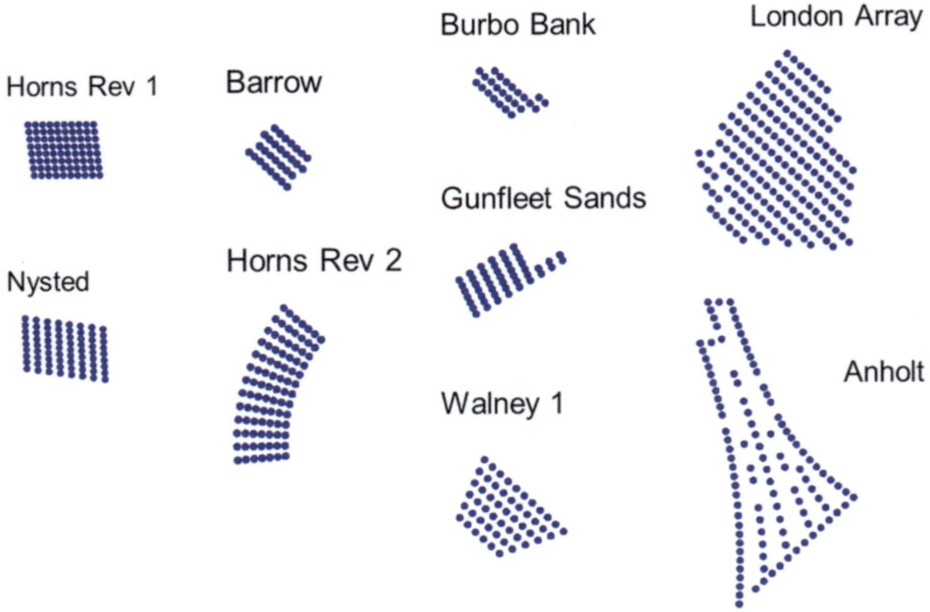


**Fig. 2.5** Array arrangement of some renowned offshore wind farms [2]

Staggered layout—turbines in alternating rows are offset (usually by half the spacing) so that no turbine is directly behind another. This triangular or staggered grid pattern exposes downstream turbines to only part of the wake from those upwind. Staggering generally increases overall production by reducing wake overlap—studies have found staggered layouts boost annual energy production by about 1–9% compared to aligned layouts. Many modern offshore farms adopt staggered patterns to enhance wake recovery and energy yield (A potential trade-off is that a turbine in a staggered layout might encounter non-uniform wind across its rotor—one side in a wake, the other in free flow—causing uneven loading and stress on blades, though this is usually manageable with robust design).

In practice, developers choose array layouts based on wind conditions and practical considerations. If a site has a strongly prevailing wind direction, turbines may be oriented in fewer rows facing that wind (to minimize depth) and spaced more closely along each row than between rows. This means the spacing in the primary wind direction is larger, while lateral spacing can be slightly tighter (since winds from the side are less frequent). In contrast, a site with multidirectional winds may favor a more symmetric or dispersed arrangement. For example, we could use a staggered/grid layout when one or two wind directions dominate, but a more matrix-like or even circular distribution if winds come evenly from many directions. Overall, most offshore wind farms employ a regular array layout (for ease of maintenance and cabling) that is refined (aligned vs. staggered, orientation of rows) to suit the wind climate and to reduce wake interference.

Several prominent offshore wind farms around the world illustrate different strategies in array layout, balancing wake minimization, cable efficiency, and site constraints. Hornsea 2, located in the North Sea off the coast of the UK, is currently the world's largest operational offshore wind farm. It uses a predominantly staggered array layout to maximize energy yield by reducing wake interference. Turbines are spaced approximately 1 km apart and offset between rows to allow better wake recovery. This layout helps maintain high capacity factors even under varying wind directions typical of North Sea conditions. Borssele 1 & 2 in the Dutch sector of the North Sea also apply a staggered grid configuration, oriented based on prevailing wind data. The developers employed computational fluid dynamics (CFD) and mesoscale atmospheric modeling to optimize the spacing—about 800–1000 m between turbines in the dominant wind direction and slightly less in the lateral direction—to reduce energy loss from wakes while keeping cable lengths and costs manageable. Walney Extension, another major UK project in the Irish Sea, features a hybrid array layout. The farm combines turbines from different manufacturers (Siemens and MHI Vestas), which are installed in distinct clusters. Within each cluster, the layout is partially aligned, optimized for the region's strong and consistent wind patterns, while maintaining clear corridors for vessel access and cable routing. Lillgrund, off the coast of Sweden, presents a contrasting case. Due to limited seabed area, it uses a tightly packed aligned layout with turbines spaced just ~ 4.3 rotor diameters apart. While this maximized installed capacity within a constrained footprint, it resulted in significant wake effects and lower energy yield per turbine. The project is often cited in literature as a cautionary example of over-compact design. These examples demonstrate how array layout decisions depend on a range of factors: wind climate, turbine size, seabed constraints, environmental regulations, and economic goals. As turbine sizes increase and digital tools improve, future wind farms are expected to adopt more site-specific, data-driven layouts, including dynamic spacing or even adaptive wake steering, to enhance both performance and ecological integration.

### 2.2.2 Wake Effects

Each wind turbine extracts energy from the wind, leaving an aerodynamic wake of slower, more turbulent air streaming behind it, as illustrated in Fig. 2.6. Turbine wakes are characterized by a velocity deficit—typically between 15 and 25% even ~ 6.5 rotor diameters (6.5 D) [3]. These wake effects cumulate in large wind farms: under stable atmospheric conditions, a cluster of turbines can create a combined wake extending 50 km downwind [4]. Wake recovery is the process by which the slowed air in a wake re-accelerates toward the ambient wind speed as it mixes with surrounding air. Recovery occurs gradually with downwind distance: significant wake losses tend to dissipate after on the order of 7–10 rotor diameters in typical conditions, though partial deficits can persist much further. Wind speed may recover to ~ 95% of the free-stream value by about 50–60 rotor diameters downstream [5]. Faster recovery is aided by turbulence—higher turbulence promotes mixing of faster-moving air into the wake, thereby restoring wind speed sooner. The turbulence intensity within wakes is generally elevated due to wind shear and vortices shed from turbine blades. This wake-added turbulence causes greater fatigue loading on downstream turbine components and support structures, even as it helps the wind speed to bounce back.

Wind direction is a critical factor in wake interactions. Wakes primarily trail downstream along the wind's path—so a turbine directly downwind of another will experience the strongest deficit. In an aligned layout, when the wind blows along the row, multiple turbines in line can be affected (a "wake shadow" through the array). By contrast, if the wind comes at an angle or if the layout is staggered, a given wake may pass beside the next turbine rather than hitting it head-on, lessening impact. To reduce wake overlap under real, shifting wind directions, developers analyze wind roses and tailor layouts accordingly. Some projects now use yaw-based wake steering—intentionally misaligning the rotor to deflect the wake away from downstream units—as demonstrated by the Wake Adapt project [6]. However, because real wind directions vary, even a well-oriented farm will see some periods when winds blow from sub-optimal angles, causing wake interactions between turbines that weren't intended to be in each other's lee. Design studies suggest tailoring the layout to the site's wind rose: for example, using staggered rows oriented for the dominant winds, or a more distributed grid if winds come from many

**Fig. 2.6** illustration of wake effects within an offshore wind farm

directions over the year. The goal is to minimize the number of turbines directly "downwind" of others in the frequent wind directions. In operation, advanced strategies like yaw control (deliberately misaligning turbines to deflect wakes) are also being explored to reduce wake losses in aligned configurations, but the fundamental prevention of excessive wake overlap starts with the array arrangement.

How offshore wind turbines are positioned relative to one another plays a critical role in determining the structural loads they experience throughout their operational life. When a turbine operates in the aerodynamic wake of another, it is subjected to reduced wind speed and elevated turbulence, both of which induce unsteady aerodynamic loading on the rotor blades and downstream structures. This wake-induced flow—characterized by velocity deficits, shear gradients, and vorticity—creates larger fatigue loads on the blades, drivetrain components, yaw systems, and tower foundations than turbines experiencing free-stream wind. A poorly arranged layout, where turbines are frequently aligned with each other under dominant wind conditions, causes certain units to be consistently exposed to turbulent inflow, accelerating material fatigue and reducing their expected operational lifespan. Conversely, a well-optimized array layout helps distribute these loads more evenly. One of the most effective strategies is the use of a staggered configuration, in which turbines are offset in alternate rows rather than directly aligned. This design ensures that no single turbine consistently operates in the direct wake of another, particularly in the prevailing wind direction. As a result, turbulent loading is distributed more equitably, reducing peak stress cycles and prolonging the structural health of all units.

Empirical data and high-fidelity modeling studies support this strategy. For instance, an optimization study found that by increasing inter-row spacing and staggering turbine orientation, it was possible to reduce monopile fatigue loads by up to 18% and cut wake interference losses by approximately 25% across the wind farm [7]. Analyses of large offshore farms link wake meandering and added turbulence to elevated fatigue-equivalent loads on downstream turbines [8]. However, layout choices also involve trade-offs. A staggered array, while beneficial in dispersing wake effects, may cause asymmetric inflow conditions, where a single rotor blade enters the wake before the others—potentially causing cyclic pitch moment imbalances. Fortunately, modern turbine control systems, such as individual pitch control (IPC) and wake-aware yaw optimization, are increasingly equipped to compensate for such transient loads [9].

### 2.2.3 Optimal Spacing Strategies

Choosing the right spacing between turbines is a balancing act between energy performance and practical constraints. If turbines are too closely spaced, wake interference will significantly lower the farm's energy output and increase mechanical stress. If spaced very far apart, wake losses become negligible—but the farm will occupy a larger area, requiring longer cables and higher installation costs for only marginal gains in energy. In offshore wind industry practice, a common rule-of-thumb is on the order of 6–10 rotor diameters separation in the prevailing wind direction. Spacing turbines about 3–5 *D* apart in the crosswind direction [10] and 6–10 *D* apart in the downwind (prevailing wind) direction can be a baseline to reduce wake overlap [11]. For instance, for a modern large turbine with a rotor ~ 180–200 m across, 6–10 *D* translates to roughly 1–2 km spacing between turbines along the dominant wind axis.

Notably, optimal spacing is not one-size-fits-all—it depends on turbine size, site wind regime, and economic tradeoffs. As turbine rotor diameters have grown (now exceeding 200 m on the newest 12–15 MW turbines), spacing in absolute meters has also increased to maintain similar *D*-based separation. A larger turbine creates a larger wake, so developers often widen the spacing to preserve performance, which lowers the power density (MW per $km^2$) of the wind farm. For example, newer European projects with 10 + MW turbines might use 10 *D* or more distance in row spacing. In contrast, some older farms or space-constrained projects were built with tighter spacing (on the order of 4–5 *D*), accepting higher wake losses. A classic case is the Lillgrund offshore wind farm in Sweden, which used only ~ 4.3 *D* spacing (due to a small site), resulting in estimated wake energy losses exceeding 20% [12].

Generally, when sea space is limited or a developer aims to maximize installed capacity in a given area (e.g. due to lease area limits or cost of additional seabed), they may pack turbines closer together, whereas a project prioritizing highest energy yield per turbine will opt for wider spacing. Research optimizations often find diminishing returns beyond a certain distance—one study suggested an optimal ~ 14.5 *D* spacing in the main wind direction (with ~ 11 *D* crosswind) to maximize yield. Such a large spacing greatly reduces wake interactions but uses a lot of ocean area [13]. In practice, most developers won't space turbines 14 *D* apart unless area is plentiful, because of the higher infrastructure cost of cabling many widely spaced units. Instead, the typical strategy is to use moderate spacing (perhaps 6–8 *D*) and a staggered layout to get most of the energy benefit of reduced wakes, without an excessive footprint.

It's also important that spacing strategies consider the prevailing wind and secondary wind directions. Often the distance between rows (in the dominant wind direction) is made larger, while the spacing along a row (side-by-side) can be slightly smaller, reflecting that winds blowing along the row will generate wake one behind another. For instance, an idealized staggered layout for a site with a single prevailing wind might have turbines ~ 7 *D* apart in the wind direction and ~ 4 *D* apart laterally. This kind of rectangular spacing (wider downwind than crosswind) is evident in many operational farms. Ultimately, the chosen spacing is a result of optimization: developers use flow models and cost models to maximize the annual energy production and profit for the farm. The spacing that maximizes energy output per turbine might not maximize output per dollar or per square kilometre—so final layouts tend to be a compromise between energy efficiency and economic density. As a result, offshore wind farm designs strike a balance where wake losses are reduced to acceptable levels (perhaps 5–15% energy loss due to wakes across the farm) while keeping the array as compact as feasible. Industry experience and studies (both computational and actual performance data) guide these decisions: for example, it's observed that beyond roughly 7–9 $D$ of separation, additional gains are small since most wake deficits have dissipated. Thus, many projects cluster around that spacing. If new technology or cost changes allow, future farms may expand spacing further—especially as very large (15 + MW) turbines demand more room aerodynamically.

### 2.2.4 Marine Ecosystem Impacts

Offshore wind farms represent a large-scale intervention in the marine environment, with both positive and negative ecological consequences. The physical footprint and spacing of turbines can significantly alter the local seascape. Each turbine foundation—whether monopile, jacket, or gravity-based—along with its surrounding scour protection (typically rock or concrete) acts as an artificial reef, attracting marine organisms to colonize the newly available hard substrate. Over time, these structures may enhance local biodiversity by offering habitat to fish, crustaceans, mollusks, and other benthic species, particularly in sediment-dominated regions where hard substrate is otherwise rare. However, construction activities can disrupt benthic communities. Pile driving, dredging, and anchoring disturb or destroy seafloor habitats, especially sensitive areas like eelgrass beds, soft corals, or spawning grounds. The impact is not limited to the installation phase—foundations alter hydrodynamic conditions by changing local current velocities and sediment transport, which can affect larval dispersal, nutrient cycling, and the structure of benthic ecosystems. Floating offshore wind farms, though often touted as less invasive, still require anchor systems and dynamic cables, which can cause localized scouring and habitat compression on the seafloor.

The spacing between turbines also influences the extent and nature of ecological impacts. A more widely spaced farm distributes its artificial reefs over a broader area, which can increase the ecological benefit by expanding habitat availability across the seabed. However, this also enlarges the zone disturbed during installation and may fragment benthic habitats. Conversely, denser layouts concentrate the impact in a smaller area but may intensify effects on water movement and ecological connectivity. Additionally, very large wind farms can subtly alter air–sea interactions: clusters of turbines extracting wind momentum can reduce turbulence and wind-driven mixing at the ocean surface in their lee, leading to minor changes in local water temperature or stratification—effects that may influence plankton productivity or larval drift. These fine-scale environmental effects, particularly how different layout geometries (e.g., long aligned rows vs. scattered arrays) influence ocean processes, are still the subject of active research.

Subsea high-voltage export cables, essential for transporting electricity back to shore, introduce additional ecological concerns. These cables emit electromagnetic fields (EMFs), which have been shown to affect behavior in marine species sensitive to electric and magnetic cues—such as sharks, skates, rays, eels, and certain migratory fish. Studies have observed that EMFs can disrupt orientation, navigation, or feeding behavior in laboratory and field settings. While EMF strength drops rapidly with distance from the cable, the cumulative impact of multiple cables crossing an area may be significant, particularly in narrow migratory corridors or estuarine environments.

Seabirds are also impacted by offshore wind development. Birds flying at rotor height—particularly species such as gannets, kittiwakes, and some gulls—face collision risks, especially during poor visibility or when attracted by fish aggregating near foundations. Furthermore, birds may be displaced from critical foraging or resting areas due to the presence and noise of turbines. Lighting on turbines, while important for navigation safety, can disorient nocturnally migrating species, compounding mortality risk. However, these impacts are highly species- and site-specific; while some seabirds may avoid turbine arrays entirely, others may adjust flight paths or exploit new foraging opportunities near foundations.

The space between turbines, while necessary for technical reasons (e.g. wake reduction, maintenance access), also translates into consumed ocean space, which can conflict with other marine users such as fishers, shipping operators, and conservationists. While wind farms often become no-fishing zones for safety reasons, this can be ecologically beneficial. Restrictions on bottom trawling and other extractive activities reduce direct seabed disturbance and create de facto marine protected areas, allowing fish stocks and benthic habitats to recover. In some European wind farms, increases in fish biomass and benthic diversity have been recorded following the introduction of no-fishing restrictions.

To mitigate these environmental impacts, a range of measures can be implemented throughout the planning, construction, and operational phases of a project. Careful site selection is perhaps the most critical step—by avoiding ecologically sensitive areas such as bird flyways, marine mammal habitats, or essential fish habitats, developers can prevent major conflicts. Prior to construction, developers conduct Environmental Impact Assessments (EIAs) and ecological baseline surveys to identify and map vulnerable areas. During installation, low-noise foundation techniques such as vibratory pile driving or suction bucket foundations can be used in place of conventional impact piling, thereby reducing acoustic disturbance to marine mammals. If impact piling is necessary, noise mitigation tools such as bubble curtains or acoustic deterrent devices can help minimize harm.

Burial of subsea cables, ideally to depths that fully attenuate EMFs, is a standard mitigation for electromagnetic impacts on fish. Additionally, cable routing can be designed to avoid sensitive habitats or be aligned with existing infrastructure corridors to reduce new disturbance. For seabirds, adaptive turbine lighting systems, which dim or flash rather than continuously shine, can reduce disorientation. In high-risk areas, temporary shutdowns during peak bird migration periods may also be warranted. Furthermore, layout design can support ecosystem function—for example, by maintaining open corridors within turbine arrays to allow movement of marine mammals, birds, and fish, and by clustering turbines in ways that minimize fragmentation of seafloor habitats.

### 2.2.5 Maintenance and Access

Offshore wind turbines are exposed to some of the harshest operating environments on Earth—strong winds, salt spray, and variable sea states—making maintenance and accessibility critical to their long-term performance and cost-efficiency. Modern offshore turbines are typically designed for a 20–25 year operational lifetime, during which they must remain not only structurally sound but also operationally efficient. However, they are far from maintenance-free. Industry benchmarking indicate that offshore turbines experience dozens of forced outages annually (despite major component repairs are infrequent), often resulting in extended downtime depending on accessibility and weather [14, 15]. Overall, the average downtime per turbine per year can be around 15 days, influenced by seasonal weather patterns and logistical delays [16].

One key form of maintenance is preventive maintenance, which involves planned, scheduled activities to ensure turbines remain in working order. This includes routine checks, blade cleaning, lubrication, bolt torqueing, and replacement of consumable parts such as filters or hydraulic fluids. These interventions are often performed annually or semi-annually and are vital for preventing premature wear and minimizing unexpected failures. Offshore conditions make this more challenging than onshore, requiring precise coordination of technicians, vessels, and equipment during short weather windows.

When turbines experience unexpected breakdowns—such as failures in gearboxes, generators, or control systems—corrective maintenance becomes necessary. These unplanned repairs can be urgent, especially if a fault halts energy production. Depending on the severity, the turbine might be shut down until technicians can reach it and restore functionality. In harsh offshore conditions, weather-related access limitations may delay these repairs for days or even weeks, increasing energy losses and operational costs.

Increasingly, wind farm operators are turning to condition-based maintenance (CBM) strategies, which use real-time data from onboard sensors to predict faults before they escalate. These systems monitor variables like vibration, temperature, and oil quality in key components, allowing maintenance teams to intervene just before a failure occurs. This predictive approach, often backed by AI-based diagnostics, has been shown to reduce downtime and extend component lifespan, ultimately lowering the levelized cost of energy (LCOE).

Accessing turbines for maintenance is a logistical challenge that depends heavily on array layout and spacing. Crew Transfer Vessels (CTVs) are the workhorses of the industry for nearshore farms. These agile boats carry up to two dozen technicians from port to turbine and are equipped to "push on" and allow safe technician transfers via ladder systems mounted on the turbine platforms. CTVs are most effective in sea states with up to around 2 m significant wave height [17], above which access becomes unsafe or impossible. For larger or more remote wind farms, Service Operation Vessels (SOVs) are deployed. These larger ships can remain offshore for weeks, housing dozens of personnel and offering motion-compensated gangways that allow safe turbine boarding even in higher waves.

More complex interventions—such as replacing blades, generators, or main shafts—require specialized jack-up vessels or floating crane barges. These ships stabilize themselves on seabed legs and lift their hulls above the waves, forming a secure base for lifting operations. These vessels are expensive and must be carefully scheduled during limited weather windows to avoid costly delays.

Array layout is carefully designed to facilitate this access. Wind farms are commonly arranged in grid-like patterns with consistent spacing—often 500–1000 m between turbines—to allow CTVs, SOVs, and crane ships to navigate safely. Straight rows aligned with cardinal or prevailing wind directions create clear navigational corridors for vessels and improve the efficiency of inspection rounds. This was a major factor in the layout of the London Array wind farm, which prioritized maintenance efficiency alongside aerodynamic performance [18].

To reduce reliance on manned access, the industry is increasingly adopting automated and remote maintenance technologies. Drones are now widely used for blade inspections, capturing high-resolution and thermal images to identify cracks, lightning damage, or erosion. These drones reduce technician risk and inspection time. Simultaneously, Remotely Operated Vehicles (ROVs) and autonomous underwater robots are being deployed for subsea inspections of foundations, cables, and scour protection. Tethered or untethered, these machines can detect corrosion, monitor cable burial depth, and assess biofouling without diver intervention. Some future-facing projects envision robotic crawlers that remain on the turbine and conduct routine maintenance tasks autonomously, such as bolt torque checks or small repairs.

## 2.3 Offshore Wind Around the World

Globally, the operational offshore wind fleet stands at approximately 83 GW as of end-2024. Asia (mainly China) leads with approximately 38–41 GW, representing 40–50% of global capacity. Europe follows closely with around 37 GW, accounting for over 40%, anchored by the UK, Germany, and the Netherlands, and anchored by the North Sea supply-chain hub. North America, led by the United States, contributes around 0.2 GW. These figures are growing rapidly, with around 170 GW of additional capacity expected to be online by 2030 under current development pipelines.

Europe has been the pioneer region for offshore wind, building out the majority of early capacity and developing mature port infrastructure, industrial supply chains, and policy frameworks. The United Kingdom, in particular, stands at the forefront, with projects like Hornsea 2 and Dogger Bank. Denmark, Germany, and the Netherlands have also developed extensive offshore networks, supported by strong government policies, port infrastructure, and industrial supply chains. The North Sea has become the epicenter of offshore wind development, with increasingly large projects being built farther from shore, often using floating substation technology and high-voltage direct current (HVDC) transmission systems to bring power back to land efficiently.

Located approximately 89 km off the Yorkshire coast in the North Sea, Hornsea 2 exemplifies the trend of building large-scale wind farms far from shore, delivering over 1.3 GW of capacity. Even larger in ambition is the Dogger Bank Wind Farm, which is being constructed in multiple phases (A, B, and C) and will ultimately reach a capacity of 3.6 GW. Once fully operational, Dogger Bank will become the largest offshore wind farm in the world and will feature GE's Haliade-X turbines, among the most powerful ever deployed. The Walney Extension, also in the UK, was previously the world's largest offshore wind farm. With a capacity of 659 MW, it is notable for using turbines from both Siemens and Vestas, showcasing interoperability in large-scale projects. In the Netherlands, the Borssele Wind Farm Zone is another significant development, comprising several projects (Borssele 1 through 5), with initial phases contributing over 750 MW and the full zone expected to exceed 1 GW. This zone has become a model for efficient permitting, fast-track development, and strong public–private coordination.

Asia is emerging as the next offshore wind powerhouse. China has surpassed all other countries in annual offshore wind installations, driven by its push to decarbonize and meet rising electricity demand in coastal provinces. With government support and a growing domestic supply chain, China's offshore capacity has grown rapidly in areas like Jiangsu and Guangdong. Projects such as the Jiangsu Rudong Offshore Wind Farm, have helped jump-start the country's coastal wind sector. Another standout is the Fujian Xinghua Bay II Demonstration Project, which brings together a range of turbine technologies, both domestic and international, in a single development zone. Meanwhile, Taiwan is making significant strides with its Formosa series of wind farms. Formosa I (128 MW), already operational, and Formosa II, under development, together represent Taiwan's commitment to building a modern offshore wind industry, targeting energy independence and industrial growth. Taiwan has become a regional leader with an aggressive rollout of fixed-bottom offshore wind farms and a strong emphasis on local content. Japan and South Korea, constrained by deep coastal waters, are investing in floating wind technology and demonstration projects that will set the stage for broader deployment in the 2030s.

In the United States, offshore wind development has been slower but is now gaining momentum. The East Coast, with its relatively shallow waters and proximity to major population centers, is home to the first commercial-scale projects such as Vineyard Wind and South Fork Wind. Federal and state-level support, including leasing auctions, tax incentives, and renewable portfolio standards, are driving forward a pipeline of over 80 GW in planned capacity. On the West Coast, floating wind is key due to deep seabeds. California and Oregon are exploring large-scale deployment. Vineyard Wind 1, located off the coast of Massachusetts, is the United States' first full-scale offshore wind farm under construction, with a planned capacity of 806 MW. It marks a major milestone for clean energy development in North America. On a smaller scale, South Fork Wind, located off New York, became the first project to deliver power from federal U.S. waters in 2024, producing 132 MW.

Beyond the traditional markets, countries like Brazil, Vietnam, and India are exploring offshore wind potential. Brazil's extensive coastline and wind resources offer significant promise, especially as technology matures and costs fall. In Southeast Asia, Vietnam's shallow southern waters are well-suited to fixed-bottom installations, while India's Gujarat and Tamil Nadu coasts have been identified as strategic zones.

Floating offshore wind is also seeing important milestones. As of early 2025, floating offshore wind remains a nascent but strategically important segment of the global offshore wind market, with approximately 275 MW installed—less than 0.5% of total offshore capacity worldwide. The United Kingdom leads the field with pioneering projects like Hywind Scotland and Kincardine, together contributing around 80 MW. Norway follows with the Hywind Tampen project, which uniquely supplies power to offshore oil platforms and adds another 88 MW. Portugal's WindFloat Atlantic contributes 25 MW, showcasing the viability of semi-submersible floating foundations. Other countries, including France, Japan, South Korea, and the United States, have launched small-scale demonstration projects. While current capacity is modest, floating wind is poised for significant growth, especially in regions with deep coastal waters unsuitable for fixed-bottom turbines.

## 2.4 Key Challenges

### 2.4.1 Capital Investment and Financial Risks

The development of offshore wind farms represents one of the most capital-intensive infrastructure undertakings in the energy sector. This high cost profile arises from both the scale of initial capital expenditure (CAPEX) and the ongoing operational expenditure (OPEX) needed to maintain reliable power generation in harsh marine environments. Offshore wind projects require billions of dollars upfront, driven by specialized equipment, installation challenges, and complex logistics.

Offshore turbines themselves are significantly larger and more expensive than their onshore counterparts. For example, hardware of a single 15 MW turbine can cost tens of millions of dollars, excluding installation costs. Foundations add another layer of expense; fixed-bottom structures like monopiles or jackets must be installed in difficult seabed conditions, requiring specialized vessels and extensive geotechnical work. Floating turbines add complexity through mooring systems and anchors.

A particularly costly component is the subsea cabling infrastructure. Inter-array cables connecting turbines to offshore substations, as well as export cables linking substations to onshore grids, represent approximately 10% of total CAPEX. These cables are not only expensive to manufacture and install but also extremely costly to repair. Marine power cables typically cost around $1–2 million USD per km, depending on voltage, cable type, and seabed conditions [19]. A cable fault often requires mobilizing expensive vessels and expert crews, with repairs sometimes taking weeks or months. This makes subsea cables one of the most financially sensitive elements of offshore wind farms.

To illustrate the scale of investment, the Hornsea 2 offshore wind farm in the UK values at about $3.7 billion USD for 1.32 GW of capacity. This translates to roughly $2.8 million per MW [20]. By comparison, modern onshore wind or solar PV projects typically cost between $1 million and $2 million per MW, highlighting the significantly higher capital intensity offshore.

Beyond CAPEX, the long-term operational expenditure is substantial due to the difficult access and harsh conditions offshore. Maintenance costs are elevated because vessels or helicopters are needed to reach turbines, weather can delay service, and components such as blades, gearboxes, and cables face accelerated wear and tear. Annual OPEX for offshore wind is around $135 per kW-year, compared to $43 per kW-year for onshore wind [21]. However, advancements in digital monitoring, automation, and predictive maintenance are gradually helping to reduce these costs.

The financial risk profile of offshore wind is marked by long payback periods due to the large upfront investment and typical project lifetimes of 20 to 25 years. Revenue uncertainty arises from energy price volatility, development delays, weather disruptions, and technical faults such as turbine downtime or cable failures. Developers often mitigate these risks through power purchase agreements (PPAs), contracts for difference (CfDs), or feed-in tariffs, although such support mechanisms are increasingly scarce. Consortium financing is common, spreading risk among utilities, pension funds, and sovereign wealth funds.

Insurance is a crucial but often overlooked element in the development and operation of offshore wind farms. It is both expensive and, increasingly, difficult to obtain. The harsh marine environment and the technical complexity of offshore wind infrastructure expose projects to a wide range of risks—from weather-related delays to costly equipment failures—which drives up insurance premiums and limits availability, especially for smaller or less experienced developers.

During the construction phase, developers typically purchase Construction All-Risk (CAR) insurance, which covers damage to turbines, foundations, cables, and other components during transport and installation. It also includes protection against third-party liability and weather-related delays. This type of coverage is particularly costly because most claims tend to arise during construction, when risks are highest and uncertainties are greatest. Premiums for CAR insurance can exceed 1.5–2% of total capital expenditure, a significant sum for multi-billion-dollar projects. These faults can take weeks or even months to locate and repair, especially in deep water or during winter seasons, leading to prolonged turbine downtime and revenue loss. As a result, insurers have become more cautious, often imposing higher premiums, stricter deductibles, or even excluding certain risks like cable failures unless stringent monitoring and trenching systems are in place.

Once a wind farm becomes operational, the insurance focus shifts to Operational All-Risk policies. These policies cover mechanical breakdowns, turbine failures, lightning strikes, fires, and—most importantly—subsea cable faults. Cable damage is by far the most frequent and expensive insurance claim in offshore wind. Cable damage accounts for over 80% of offshore wind insurance claims [22].

For many projects, securing comprehensive insurance is no longer straightforward. Developers working with innovative technologies—such as floating wind platforms or new mooring systems—often find it harder to obtain coverage due to the lack of long-term performance data. Projects located in extreme environments, such as deepwater or typhoon-prone regions, face similar hurdles. Even experienced developers are seeing higher deductibles, longer claim processes, and more stringent underwriting requirements. Some insurers are withdrawing from the offshore wind sector altogether or capping their exposure to limit potential losses.

The rising cost and complexity of insurance directly affect project financing. Banks and institutional investors rely on insurance to de-risk projects. When insurance is expensive or limited, lenders may impose stricter terms, demand higher interest rates, or reduce their loan exposure. Some projects have even faced delays in financial close because they could not secure the required insurance in time. This is particularly challenging for smaller or first-time developers, who may lack the track record or capital reserves to absorb higher premiums or self-insure minor risks.

To respond to this challenge, developers are adapting in several ways. Some are investing in better digital monitoring and predictive maintenance systems to demonstrate lower risk profiles to insurers. Others are forming risk-sharing pools, particularly among state-backed utilities, to collectively manage losses. A few large players are exploring self-insurance for smaller or more manageable risks. Meanwhile, some governments and multilateral institutions—such as the European Investment Bank (EIB)—are considering insurance guarantees or other financial instruments to help de-risk offshore wind investments and promote wider industry participation.

Political and regulatory barriers add another layer of complexity to offshore wind development. Projects often face permitting delays due to lengthy environmental assessments, public opposition, and conflicts with existing marine activities such as shipping, fishing, or naval operations. Grid connection uncertainty—especially in regions without coordinated transmission planning—can further delay project commissioning. While countries like the UK and Denmark have streamlined their permitting processes through centralized auctions and clear regulatory frameworks, progress in the United States has been more uneven. Although the U.S. has significant offshore wind potential, development has been hindered by fragmented jurisdictional authority, litigation related to wildlife and coastal impacts, and shifting political priorities. In recent years, federal support for offshore wind has weakened amid a broader policy emphasis on protecting domestic oil and gas interests. This political backdrop has made long-term investment riskier and less attractive, contributing to project delays and cancellations despite earlier momentum.

### 2.4.2 Condition Monitoring and Structure Health Assessment

In offshore wind, the ability to monitor structural and operational health is as critical as the ability to access and repair turbines. Whilst it is necessary maintenance and access to enable the physical execution of repairs—vessels, crew transfers, weather windows—there is still a need to develop a monitoring and assessment framework focuses on how issues are detected, diagnosed, and pre-empted before they require costly intervention.

Condition monitoring (CM) systems are now a standard component of modern offshore wind farms. They provide continuous, real-time data on key parameters such as blade root strain, tower and foundation vibration, gearbox oil condition, generator temperature, yaw misalignment, and even nacelle motion in floating platforms. Data are typically collected via networks of accelerometers, strain gauges, fiber-optic Bragg grating sensors, corrosion probes, and LIDAR-based wind scanners. CM technologies allow operators to predict faults before they cause downtime, reducing operational expenditure and extending turbine life.

The benefits are substantial: industry studies indicate that CM and predictive maintenance can reduce offshore wind unplanned O&M costs and increase turbine availability. For example, one study found up to an 8% reduction in direct O&M costs and an 11% drop in lost production when predictive monitoring of key components was used [23]. While many farms already operate at availability levels of ~94–95%, enhanced monitoring may help push that figure higher. This translates to millions in avoided lost revenue over a farm's lifetime, especially given that a single day of downtime for a 15 MW turbine at a high-wind site can mean > $20,000 in lost electricity sales.

Emerging technologies are extending CM beyond turbine nacelles. Digital twins—virtual replicas of physical assets—are now deployed at the wind farm level. They can simulate fatigue life under site-specific wind/wave climates, assess wake-induced loads across the array, and evaluate the impact of control changes before implementing them in the field. For example, Ørsted has implemented drone-based inspections and AI analytics at its West of Duddon Sands site [24].

Despite these advances, the offshore environment imposes severe challenges on CM systems themselves. Sensors must survive constant exposure to saltwater spray, high humidity, temperature swings, and mechanical shocks from wave-induced motions. Underwater sensing—critical for foundation scour monitoring, cable integrity checks, and mooring line health—faces even harsher conditions. Existing commercial sensors may be too bulky, rigid, or fragile for integration into complex subsea geometries, and traditional bonding or rivet joining can fail under cyclic loads or introduce residual stresses. Furthermore, biofouling can degrade sensor accuracy and lifespan, and replacing subsea devices at depth is costly and risky.

Data transmission from offshore assets also presents a bottleneck. While nacelle-mounted CM systems can transmit via high-bandwidth radio or microwave to shore, subsea sensors rely on fiber-optic links or low-bandwidth acoustic modems. Underwater acoustic communication is limited by attenuation, noise interference, and range. Real-time subsea data often require advanced signal processing or storage-and-forward approaches.

Addressing these barriers will require a combination of technological innovation and system integration. One priority is the development of durable, marine-grade sensor designs that are lightweight, corrosion-resistant, and resistant to biofouling, while also able to withstand the high pressures and dynamic loads encountered in offshore environments. Such designs would ensure long-term measurement reliability, even under constant exposure to saltwater, wave motion, and temperature fluctuations. Equally important is the creation of modular, easily replaceable sensing units. By enabling rapid swap-out during short weather windows, these systems can reduce vessel time and lower the costs associated with prolonged offshore interventions. Modularity also supports upgrades and retrofits as monitoring technologies evolve. In some cases, sensors could be embedded directly into modular components—such as blade shells, tower sections, or foundation elements—so that monitoring capability is an inherent part of the structure. This approach not only streamlines installation but also protects sensors from environmental damage, creating a strong synergy between structural design and condition monitoring.

While advanced sensing systems can provide continuous streams of operational and environmental data, translating these measurements into actionable insights—particularly an accurate prediction of structural lifetime—remains a significant challenge for the offshore wind industry. The core difficulty lies in bridging the gap between raw sensor data (strain, vibration, corrosion rates, temperature cycles) and the complex, multi-physics models that govern fatigue, fracture, and ultimate strength in real-world turbine structures.

Current structural integrity assessments in offshore wind often rely on design-stage fatigue analysis using simplified load spectra and idealised geometries, guided by standards such as IEC 61400-3 and DNV-ST-0126 [25, 26]. These approaches, while useful for certification, are not fully suited to in-operation prediction because they rarely incorporate the evolving material state of the structure, site-specific environmental variability, or complex load interactions over time. For example, a monopile foundation and a semi-submersible floater will experience fundamentally different combinations of bending moments, torsional loads, and localised stress concentrations—even if the environmental forcing is similar—due to differences in geometry, stiffness distribution, and dynamic behaviour.

An effective assessment framework needs to begin with high-fidelity, geometry-specific structural models that can capture localised stress responses in the complex shapes found in offshore turbines. Welded joints, bolted flanges, and composite–metal interfaces must be represented in sufficient detail to resolve local fatigue hotspots. These models form the backbone for interpreting sensor data in the context of actual structural behaviour.

Equally essential is the use of material degradation models that reflect the properties and failure modes of the actual materials used in each component—be they S355 steel, glass-fibre composites, or hybrid laminates. Such models must account for degradation under cyclic saltwater exposure, temperature variation, and load changes induced by phenomena such as biofouling, which can subtly alter hydrodynamic forces over time.

The framework should integrate sensor-driven data assimilation techniques, such as digital twin updating, where live measurements continuously refine finite element or reduced-order fatigue models. This dynamic feedback loop enables more accurate predictions of remaining useful life (RUL) by aligning theoretical models with real-world operational behaviour.

However, the current state of practice reveals a lack of robust, validated mechanical algorithms capable of processing diverse sensor inputs into accurate lifetime predictions across multiple turbine types and materials. Algorithms that work well for tubular steel monopiles may not generalise to jacket nodes, composite blades, or hybrid floating platforms, each of which has distinct failure modes and stress–strain behaviours. Furthermore, load interactions are not purely linear—blade pitch adjustments, yaw control strategies, and even wake steering can create asymmetric fatigue patterns that are difficult to model with standard S–N curve approaches.

In other engineering sectors, such as aerospace, probabilistic damage tolerance analysis has been successfully integrated with in-situ health monitoring to estimate remaining life under uncertainty. Offshore wind could benefit from similar methods, incorporating probabilistic load spectra, stochastic material properties, and Bayesian updating from sensor data. Such an approach would allow operators to quantify confidence levels in lifetime predictions rather than relying solely on deterministic fatigue safety factors.

The opportunity ahead is to develop geometry- and material-aware mechanical algorithms that evolve in parallel with digital twin platforms, producing lifetime estimates that adapt dynamically to measured behaviour. Achieving this will require close collaboration between structural engineers, materials scientists, and data scientists, as well as validation campaigns using full-scale offshore test structures. Without this integration, the offshore wind sector risks underutilising its rapidly expanding condition monitoring infrastructure—collecting vast amounts of data but still lacking the analytical capability to fully translate them into decisions that extend structural life and reduce O&M costs.

### 2.4.3 Combined Aerodynamics, Hydrodynamics and Structural Dynamics of Floating Systems

Floating wind turbines, essential for deep-water renewable energy deployment, differ fundamentally from fixed-bottom designs because their supporting platforms are free to move in six degrees of freedom (6-DOF): surge, sway, heave, roll, pitch, and yaw. These motions are driven by waves, currents, and wind loads on the entire structure, and they interact continuously with aerodynamic forces on the rotor blades. This creates a coupled aero–hydro–servo–elastic system, where platform motion influences rotor aerodynamics, and rotor aerodynamics, in turn, can excite or damp platform motion, as shown in Fig. 2.7.

The impact of this coupling is significant for both structural loading and energy yield. Pitch motion from long-period swell can cyclically alter the effective angle of attack on the blades, leading to fluctuating aerodynamic torque and drivetrain loads. Heave motion changes the rotor's position within the wind profile, affecting inflow velocity and turbulence intensity. Surge and sway can shift the rotor into or out of the wakes of neighbouring turbines, producing abrupt changes in power output. In some cases, resonant interactions between wave periods and platform natural frequencies have been shown to amplify the effects [28].

Accurately predicting this coupled behaviour is challenging because it requires simultaneously resolving two complex, interacting fluid systems—air and water—across very different physical regimes and timescales. The aerodynamic side involves three-dimensional, unsteady flow over rotating blades, influenced by yaw misalignment, dynamic stall, and turbulent inflow. The hydrodynamic side must account for nonlinear wave loading, viscous damping, mooring dynamics, and second-order wave drift forces in irregular seas. The floating platform's motion modifies rotor inflow, which alters aerodynamic loads, which then feed back into platform dynamics—a closed-loop system with strong bidirectional coupling.

Advanced Computational Fluid Dynamics (CFD) is increasingly used to resolve this complexity by directly simulating the unsteady interaction between atmospheric turbulence and wave-induced platform motion. High-fidelity CFD can capture phenomena like dynamic stall on moving blades or wave run-up on platform columns, which lower-order models miss. However, CFD simulations of full-scale floating turbines are computationally intensive, often requiring hybrid modelling approaches where CFD is applied to critical subsystems or short-duration events, while lower-fidelity codes handle long-term load simulations [29]. On the hydrodynamic side, CFD methods such as Reynolds-Averaged Navier–Stokes (RANS) or Smoothed Particle Hydrodynamics (SPH) can resolve nonlinear wave–structure interactions and mooring fairlead loads more accurately than potential flow theory. Yet, coupling these to turbine aerodynamics in a time-accurate manner remains a technical bottleneck due to differences in computational timestep requirements between air and water solvers.

The coupling challenges extend beyond turbine and platform behaviour into mooring systems and dynamic power cables [30]. Moorings—whether catenary, taut-leg, or semi-taut—not only keep the platform in position but also influence its motion through stiffness and damping characteristics. Wave-induced platform motions can introduce cyclic tension variations in mooring lines, potentially causing fatigue in chain links, synthetic ropes, or anchors. These cyclic loads may be amplified under certain wind–wave alignment

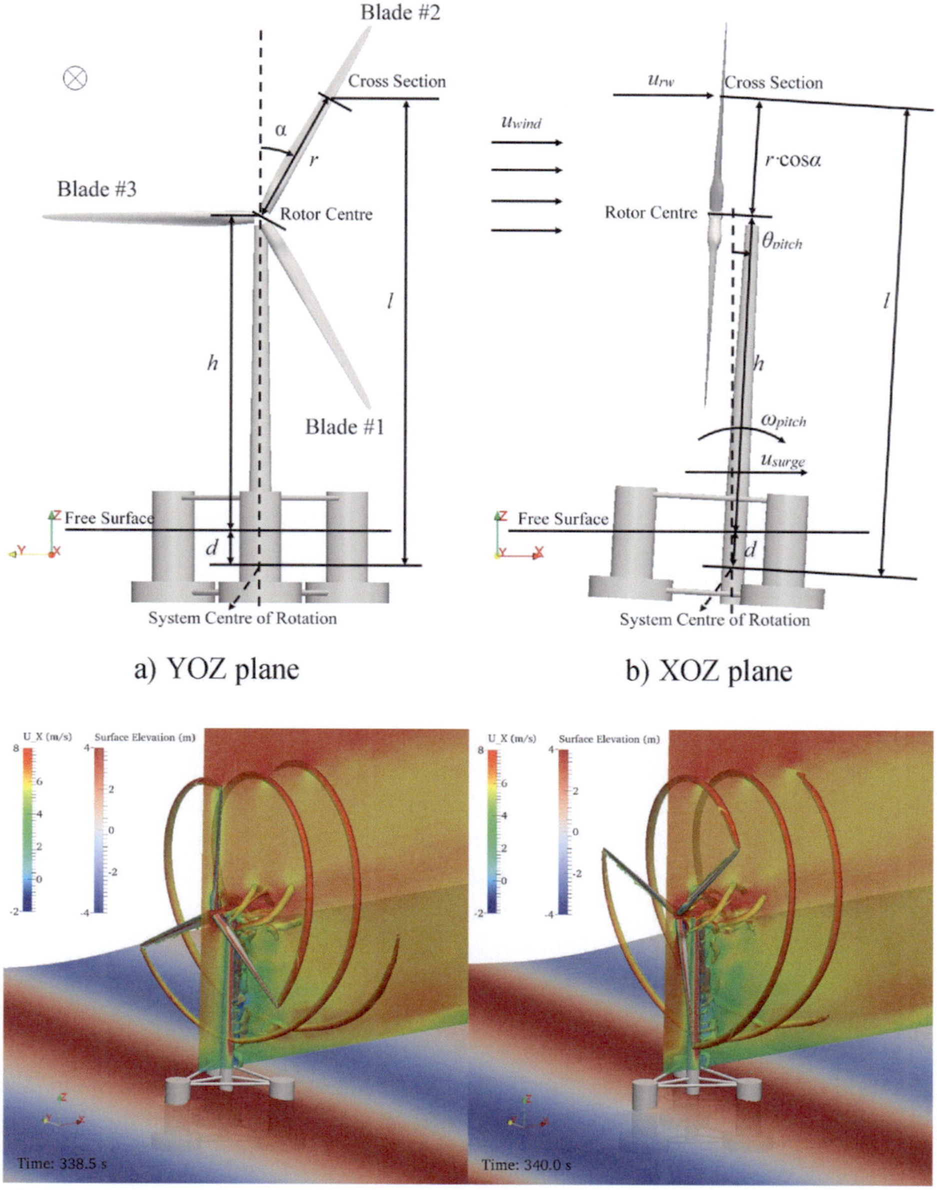


**Fig. 2.7** Illustration of combined aerodynamics and hydrodynamics of a floating wind turbine [27]

conditions, and their magnitude is strongly influenced by the platform's pitch and surge response.

Dynamic power cables face similar fatigue challenges. Their configurations—such as lazy-wave or steep-wave arrangements—are designed to accommodate the platform's 6-DOF motions without exceeding allowable bending or tensile limits. However, combined wind and wave excitation can create complex motion patterns at the touchdown point on the seabed, where bending fatigue is most critical. Coupling between aero–hydro dynamics and cable response is particularly important in long-term design, since unexpected motion–tension interactions can accelerate fatigue damage and reduce cable lifespan [31].

Finite Element Analysis (FEA) plays a key role in predicting mooring and cable fatigue life. Detailed FEA models allow engineers to assess stress distributions, bending strain ranges, and localised hot spots under realistic dynamic loading extracted from coupled aero–hydro simulations. For dynamic cables, FEA can be coupled with hydrodynamic solvers to simulate cable curvature cycles over years of operation, feeding into fatigue life calculations that account for material degradation and environmental variability.

Integrating coupled motion behaviour into structural integrity assessment frameworks is still a major challenge. Floating wind introduces load paths absent in fixed-bottom designs, such as torsional loading from yaw–surge coupling, multi-axial fatigue at the tower–platform interface, and mooring-induced torsion. These effects are inherently stochastic, driven by the joint probability distribution of turbulent wind and irregular sea states.

Here, structural FEA becomes indispensable. By importing time-domain load histories from coupled aero–hydro simulations, engineers can assess fatigue hot spots in welded joints, composite–metal transitions, and blade roots. However, without reliable input from coupled CFD–FEA workflows, lifetime predictions risk large uncertainty margins. This is compounded by the lack of validated mechanical algorithms capable of translating complex multi-axial loading histories into accurate remaining useful life (RUL) estimates across varied floating platform geometries and materials.

Bridging this gap will require multi-scale, multi-physics simulation workflows where CFD captures the aerodynamic and hydrodynamic load generation, global-scale FEA models platform and mooring response, and local FEA submodels resolve stress concentration details. Such integrated modelling, validated against full-scale measurement campaigns, will be critical to reducing uncertainty in design safety factors while ensuring reliability in long-term operation.

## References

1. Buljan A. Offshore Wind Turbines in 2024: 20+ MW Prototypes Rolling Out in Europe, China; 16 MW Remains Most Powerful Turbine Installed Offshore. OffshoreWIND.biz, 1 January 2025.

2. Nygaard N. G. Systematic quantification of wake model uncertainty. EWEA Offshore 2015, 2015.
3. Aitken M.L., Banta R.M., Pichugina Y.L. & Lundquist J.K. Quantifying wind turbine wake characteristics from scanning remote sensor data. J. Atmos. Oceanic Technol. 31, 765-787 (2014).
4. Krishnamurthy R., Newsom R.K., Kaul C.M., Letizia S., Pekour M., Hamilton N., Chand D., Flynn D., Bodini N. & Moriarty P. Observations of wind-farm wake recovery at an operating wind farm. Wind Energy Sci. 10, 361–380 (2025).
5. Dong G., Li Z., Qin J. & Yang X. How far the wake of a wind farm can persist for? Theor. Appl. Mech. Lett. 12, 100314 (2022).
6. Siemens Gamesa Renewable Energy (SGRE). Siemens Gamesa now able to actively dictate wind flow at offshore wind locations – Launch of "Wake Adapt" feature. Press release, 26 November 2019.
7. Ali S., Park H., & Lee D. Multi-Criteria Optimization of Wind Turbines in an Offshore Wind Farm with Monopile Foundation Considering Structural Integrity and Energy Generation. J. Mar. Sci. Eng. 12, 2313 (2024)
8. Moens M. & Chatelain P. Correlations Between Wake Phenomena and Fatigue Loads Within Large Wind Farms: A Large-Eddy Simulation Study. Front. Energy Res. 10, 881532 (2022).
9. Campagnolo F., Castellani F., Natili F., Astolfi D., and Mühle F. Wind Tunnel Testing of Yaw by Individual Pitch Control Applied to Wake Steering. Front. Energy Res. 10, 883889 (2022).
10. RESDM. Wind Turbine Spacing: Distance Between Turbines Explained. RESDM blog
11. Meyers J. & Meneveau C. Optimal turbine spacing in fully developed wind farm boundary layers. Wind Energy 15, 305-317 (2012).
12. Sebastiani A., Castellani F., Crasto G. & Segalini A. Data Analysis and Simulation of the Lillgrund Wind Farm. Wind Energy 24(6), 634–648 (2021).
13. Zhang X., Wang Q., Ye S., Luo K. & Fan J. Efficient layout optimization of offshore wind farm based on load surrogate model and genetic algorithm. Energy 315, 133106 (2024).
14. ORE Catapult & The Crown Estate. System Performance, Availability and Reliability Trend Analysis (SPARTA): 2020/21 Portfolio Review. June 2022.
15. Carroll J., McDonald A. & McMillan D. Failure rate, repair time and unscheduled O&M cost analysis of offshore wind turbines. Wind Energy 19, 1107-1119 (2016).
16. Sheng S. Report on Wind Turbine Subsystem Reliability – A Survey of Various Databases. NREL/PR-5000-59111. National Renewable Energy Laboratory (June 2013).
17. Phillips S., Shin I.B., & Armstrong C. Crew Transfer Vessel Performance Evaluation. In: Proceedings of the Design & Operation of Wind Farm Support Vessels Conference (London, UK, 28–29 Jan 2015).
18. https://londonarray.com/
19. McKay B., Optimised Cable Connection Options for Floating Offshore Wind. Report PN 000463–CFAR-OC-028-30092022. ORE Catapult, 30 September 2022.
20. Raymond X. Hornsea 2 Offshore Wind Farm: Project Profile – Off the coast of Yorkshire, UK. Blackridge Research, 31 October 2025.
21. Stehly T., Duffy P. & Mulas Hernando D. Cost of Wind Energy Review: 2024 Edition. National Renewable Energy Laboratory (NREL/PR-5000-91775), November 2024.
22. DNV. 80 % of insurance claims in offshore wind are related to subsea cable failures – how can the industry manage these risks? DNV, 2024.
23. Turnbull A. & Carroll J. Cost Benefit of Implementing Advanced Monitoring and Predictive Maintenance Strategies for Offshore Wind Farms. Energies 14, 4922 (2021).
24. Madhiraju P. How drone-based predictive maintenance is transforming offshore wind O&M in 2025. Business News Today, 24 June 2025.

25. International Electrotechnical Commission (IEC). IEC 61400-3-1:2019 – Wind turbines — Part 3-1: Design requirements for fixed bottom offshore wind turbines. Geneva: IEC (2019).
26. DNV. DNV-ST-0126: Support structures for wind turbines. DNV GL (2016).
27. Liu Y., Xiao Q., Incecik A., Peyrard C. & Wan D. Establishing a fully coupled CFD analysis tool for floating offshore wind turbines. Renew. Energy 112, 280–301 (2017).
28. Su O., Li Y., Li G., Cui Y., Li Y., Wang B., Hang M. & Liang J. Nonlinear harmonic resonant behaviours and bifurcation in a two degree-of-freedom Duffing oscillator coupled system of tension-leg platform type floating offshore wind turbine. Chaos Solitons Fractals 189, 115656 (2024).
29. Huang L., Li Y., Benites-Munoz D., Windt C.W., Feichtner A., Tavakoli S., Davidson J., Paredes R., Quintuna T., Ransley E., Colombo M., Li M., Cardiff P. & Tabor G. A Review on the Modelling of Wave-Structure Interactions Based on OpenFOAM. OpenFOAM® Journal 2, 116–142 (2022).
30. Wei Y., Yu S., Jin P., Huang L., Elsherbiny K. & Tezdogan T. Coupled analysis between catenary mooring and very large floating structure with structural hydroelasticity in waves. Mar. Struct. 103516 (2023).
31. Cerik B. C. & Huang L. Recent advances in mechanical analysis and design of dynamic power cables for floating offshore wind turbines. Ocean Eng. 311, 118810 (2024).

# Wave Energy Converters

# 3

Among the various ocean renewable energy technologies, wave energy converters (WECs) hold a unique and promising position due to the remarkable energy density of water. Unlike air, water is nearly 800 times denser, meaning that the power contained in ocean waves can far exceed that of wind at comparable speeds. Anyone who has stood on a beach watching the ocean waves or experienced the raw power while surfing understands intuitively the immense energy contained within the movement of water. While it is possible for a human to stand upright against strong winds, the force of crashing waves can be overwhelming and humbling, demonstrating nature's untapped power waiting to be harnessed. This intrinsic advantage suggests that wave energy has tremendous potential to contribute significantly to the global renewable energy portfolio.

In 1799, Pierre-Simon Girard filed a patent for WEC which is the first documented attempt to harness wave energy. The world, however, was distracted by war and revolution, and their idea faded into obscurity, preserved only in dusty patent records. For decades, the Girards' vision lay dormant. In the mid-1800s, inventors across Europe and America began to revisit the concept. By the 1850s, rudimentary wave-powered pumps appeared, used to desalinate water or drive small mills. The dawn of the twentieth century brought new tools: electricity, advanced materials, and a growing hunger for energy. In the 1940s, Yoshio Masuda designed the first modern WEC, a floating buoy that converted wave motion into electricity via an air turbine. By 1947, his Oscillating Water Column (OWC) device powered navigational buoys, a small but revolutionary step. In the UK, Stephen Salter's "Duck" emerged in 1974, a nodding, wave-riding device that captured energy with remarkable efficiency in lab tests. These early WECs were inefficient and struggled to survive in storms. By the early 2000s, climate change cast a long shadow. The need for clean, renewable energy revived interest in WECs. In 2008, the Pelamis,



L. Huang, *Ocean Renewable Energy Systems*, Synthesis Lectures on Renewable Energy Technologies, https://doi.org/10.1007/978-3-032-10576-9_3

a snake-like attenuator, became one of the world's first commercial-scale wave energy projects off Portugal's coast. Its sleek, red segments undulated with the waves to generate electricity, but financial and technical challenges forced its retirement. Meanwhile, companies like Ocean Power Technologies and Eco Wave Power pushed forward, deploying point absorbers and shore-based systems [1].

Despite this long history, today the industry has yet to see a widely accepted, commercially successful WEC design. The reason lies partly in a fundamental engineering dilemma: the delicate balance between structural integrity and energy efficiency. WECs must be robust enough to withstand harsh marine environments, all while maximizing energy extraction from the waves. In other words, WECs are intended to be deployed in large waves for substantial energy yield, while this also tends to induce large structural loading. Designs that prioritize efficiency often lead to fragile structures prone to failure, whereas those built for strength may be too costy or sacrifice the amount of power they can capture. This tension has kept wave energy in a challenging position, sparking ongoing innovation and debate among researchers and engineers. As of 2025, the global installed capacity remains low (on the order of tens of MW), significantly underusing the theoretical potential. This gap highlights both the promise and the challenges of the technology.

In this section, we delve deeper into the world of wave energy converters, exploring their working principles, classifications, and the key technological challenges they face. Understanding these will highlight both the immense opportunities and the persistent obstacles on the path toward unlocking the ocean's clean, renewable power.

## 3.1 Operating Principles

WECs are devices engineered to capture the kinetic and potential energy associated with ocean waves and convert it into usable forms, predominantly electricity. The process involves converting the mechanical energy of waves through their movement into electrical energy. There are many types of WECs, depending on various working mechanisms and application scenarios. Unlike wind turbines, there has not been a single prototype of WECs that can enable mass production. Thus, many types of designs are being tested and evolving.

### 3.1.1 Point Absorbers

Point absorbers are a class of WECs designed to extract energy from the motion of waves at a single point, typically through the relative movement between a floating or submerged buoy and a fixed reference structure such as the seabed or a secondary mass, as illustrated in Fig. 3.1. These devices are generally compact, symmetrical, and moored to remain stationary in the horizontal plane while oscillating up and down with passing waves.

Point absorbers work by exploiting the pressure differential or motion caused by wave peaks and troughs. As the buoy moves with the wave, it drives a Power Take-Off (PTO) system—commonly hydraulic rams, linear electric generators, or mechanical linkages—which converts mechanical motion into electricity. Depending on the design, the system can operate efficiently in a broad range of sea states and wave directions due to its omni-directional energy capture capability.

One of the main advantages of point absorbers is their omni-directional energy capture. Because they respond to vertical wave motion from any direction, they are well-suited to sites with complex or multidirectional wave climates. Their relatively small size and modularity make them attractive for deployment in arrays (known as wave farms), allowing developers to scale up generation capacity incrementally. In addition, many point absorbers float at or near the surface, simplifying installation and maintenance compared to fully submerged or seabed-anchored devices. These features make them versatile for different applications, from powering offshore sensors and autonomous vehicles to contributing electricity to coastal grids or microgrids.

However, point absorbers also face several challenges. Their continuous exposure to the harsh marine environment—including extreme wave events, corrosion, and biofouling—places considerable demands on structural design, durability, and survivability. The

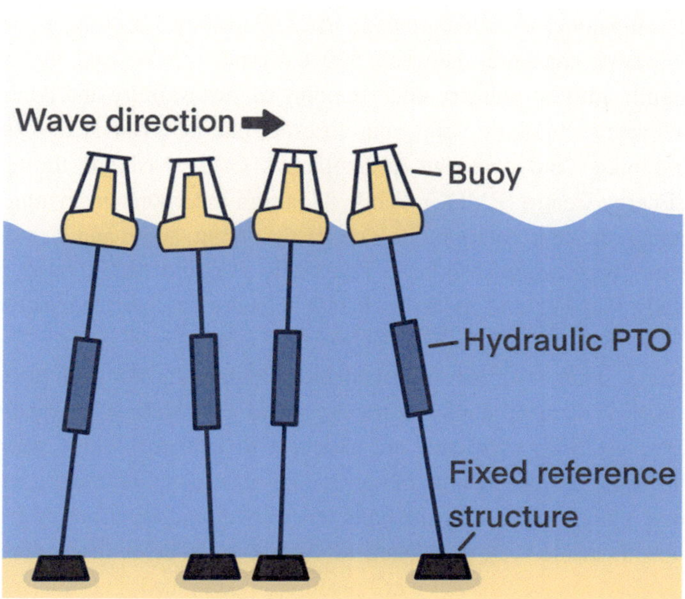


**Fig. 3.1** Illustration of point absorbers WEC

need for robust mooring and anchoring systems to maintain position without excessive tension adds further complexity and cost. Furthermore, their performance is highly sensitive to the local wave spectrum; they tend to operate most efficiently within a specific range of wave heights and periods, and can experience a sharp drop in energy capture when wave conditions deviate from the design optimum. In large arrays, hydrodynamic interference between devices can reduce individual performance due to wave shadowing effects, complicating layout optimization.

Several point absorber projects have been developed and tested worldwide, showcasing different technical approaches and scales of operation. One notable example is the CETO system developed by Carnegie Clean Energy in Australia. CETO units are fully submerged point absorbers tethered to the seabed, with each unit driving a hydraulic PTO to pump pressurized seawater to shore. The CETO 5 units demonstrated off Garden Island had a peak capacity of 240 kW per unit, and the company has since been developing larger-scale versions, such as the CETO 6, targeting 1 MW capacity per device.

In the United States, Ocean Power Technologies (OPT) has developed the PowerBuoy system, a surface-floating point absorber designed for remote ocean applications. The PB3 PowerBuoy, OPT's flagship model, is capable of producing an average of 150–350 W of continuous power, with peaks in the kilowatt range depending on site conditions. Although modest in output, the PB3 is designed for long-duration deployments to support instruments, communication systems, and subsea monitoring infrastructure.

Ireland's Wavebob project was another early example of point absorber technology, featuring a dual-body heaving system optimized for the North Atlantic wave climate. Although the company ceased operations in 2013, Wavebob's prototypes were estimated to have peak power capacities between 500 kW and 1 MW, and the data collected contributed significantly to industry understanding of survivability and control systems.

Another relevant technology with point-absorber-like behavior is the Wello Penguin, developed in Finland and deployed in Scotland's Orkney Islands through the European Marine Energy Centre (EMEC). While its design incorporates asymmetrical motion to extract energy through rotation, its compact floating body and heaving dynamics share similarities with point absorbers. The Wello Penguin had a rated power output of approximately 500 kW, and its deployments helped assess performance in real ocean conditions.

Looking ahead, point absorbers offer considerable potential for both grid-scale energy generation and niche marine applications. Advances in materials engineering, adaptive control systems, and PTO efficiency are expected to enhance device performance and longevity, while modular production and array optimization could reduce levelized costs. With increasing global interest in offshore renewables and decarbonized ocean operations, point absorbers are well-positioned to contribute to diversified energy portfolios, especially in regions with consistent wave resources. Continued investment in demonstration projects, supportive regulatory frameworks, and hybrid integration with wind or solar systems will be crucial for realizing their commercial viability.

### 3.1.2 Terminators

Terminator WECs are designed to absorb wave energy by directly facing and intercepting the direction of wave travel. Positioned perpendicular to the wave front, these devices "terminate" the wave's progression, extracting energy from the horizontal and vertical movement of the water. Terminators typically span a substantial portion of the wave front and are either fixed to the seabed, embedded in the shoreline, or installed as floating hinged structures. Among these, Oscillating Wave Surge Converters (OWSCs) represent an outstanding subclass, utilizing a bottom-mounted, pivoted flap that swings back and forth in response to the surge motion of waves in shallow waters, as shown in Fig. 3.2.

OWSCs convert the oscillatory surge of water particles near the seabed, typically found in water depths of 10–25 m, into rotational mechanical energy. As the flap swings, it drives a hydraulic power take-off system that pressurizes fluid to turn a generator. Because the surge motion is generally stronger closer to shore, OWSCs are well-suited for nearshore deployment and benefit from shorter transmission distances to grid connection points. Their ability to operate efficiently in relatively shallow waters makes them compatible with a wide range of coastal environments and reduces installation and maintenance complexity compared to deeper offshore systems.

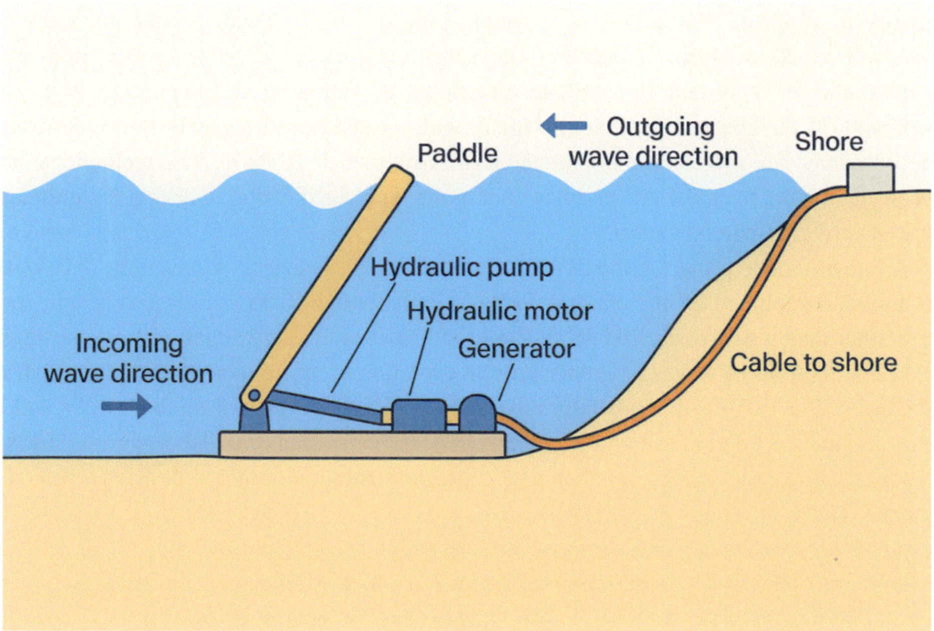


**Fig. 3.2** Illustration of an oscillating wave surge converter

One of the key advantages of terminator-type WECs, including OWSCs, is their high theoretical energy capture potential. By engaging a broad cross-section of the wave front, these devices can extract more energy than point absorbers under optimal conditions. OWSCs, in particular, offer favorable efficiency in shallow waters and often have a relatively low visual impact when only the upper portion of the flap is visible above the surface. Maintenance access is easier compared to fully submerged or deep offshore systems, and nearshore installations allow for simpler electrical infrastructure. In some cases, terminator devices can also double as coastal defense structures, absorbing wave energy that would otherwise cause erosion.

However, several challenges limit the widespread deployment of these technologies. Terminator devices must withstand significant structural loads due to direct wave impact along their entire face. This necessitates robust engineering and can drive up material and installation costs. OWSCs, though mechanically simpler than some offshore systems, still face durability issues caused by biofouling, corrosion, and fatigue from continuous oscillation. Additionally, energy output is sensitive to wave direction and period, and in areas with highly variable or multidirectional wave climates, performance may be reduced. Grid integration for nearshore or remote deployments can also present technical and regulatory hurdles, particularly in regions with weak coastal energy infrastructure.

Several demonstration projects have shown the real-world viability of terminator and OWSC technologies. A leading example is the WaveRoller system developed by AW-Energy of Finland. WaveRoller is a bottom-hinged OWSC installed near the shore in waters 8 to 20 m deep. It captures the surge component of wave motion and uses a hydraulic PTO system to generate electricity. A full-scale demonstration unit was deployed off the coast of Peniche, Portugal, with a rated power capacity of 350 kW, and the company has proposed commercial-scale arrays of 1–10 MW. The technology has received support from the EU and has undergone type certification, marking a significant step toward commercialization.

Another notable project is the ISWEC (Inertial Sea Wave Energy Converter), developed in Italy and deployed off the coast of Pantelleria. Although ISWEC uses gyroscopic control rather than a flap-based OWSC, it shares similarities in energy capture from the surge and pitch motions of waves and represents an evolution of terminator concepts for floating platforms. The ISWEC demonstrator has produced power outputs of up to 100 kW and is being scaled for offshore oil and gas platform decarbonization in collaboration with ENI.

Historically, one of the earliest and most influential terminator concepts was the Salter's Duck, invented in the 1970s. This bobbing device was designed to rotate in response to wave impact and demonstrated very high theoretical efficiencies—up to 90% in lab tests—but was never commercialized. Nonetheless, it paved the way for future terminator designs and remains a benchmark for energy capture efficiency.

In the future, terminator WECs, especially OWSCs, are poised to play a valuable role in the global marine energy mix. Their suitability for nearshore sites makes them attractive for island nations, coastal microgrids, and hybrid coastal protection systems. As materials technology improves and real-time control systems become more advanced, the reliability and efficiency of OWSCs are expected to improve. The ability to integrate energy storage and smart grid interfaces will further enhance their value. To scale up, ongoing research into array layouts, load management, and long-term environmental impact will be essential. With successful field trials and growing investor interest, OWSCs in particular are well-positioned to transition from prototypes to commercially viable deployments. As coastal regions seek clean, local energy sources, and as wave resource assessments continue to improve, terminator devices may find a niche in complementing offshore wind and solar in diverse, mixed renewable energy systems.

### 3.1.3 Attenuator

Attenuator wave energy converters are long, articulated floating structures aligned parallel to the direction of incoming waves. Rather than obstructing or terminating the wave front like terminators, attenuators ride along the surface of the waves, flexing as they follow the wave profile. This motion creates relative movement between the segments of the structure, which is then converted into electricity through hydraulic pumps, linear generators, or other power take-off (PTO) systems located at the joints, as shown in Fig. 3.3. By absorbing wave energy over an extended structure length, attenuators offer a distributed approach to energy capture, making them well-suited to offshore environments where wave energy is abundant and relatively uniform.

A well-known example of an attenuator WEC is the Pelamis Wave Energy Converter, developed in the UK. Pelamis consisted of several cylindrical segments, each about 30

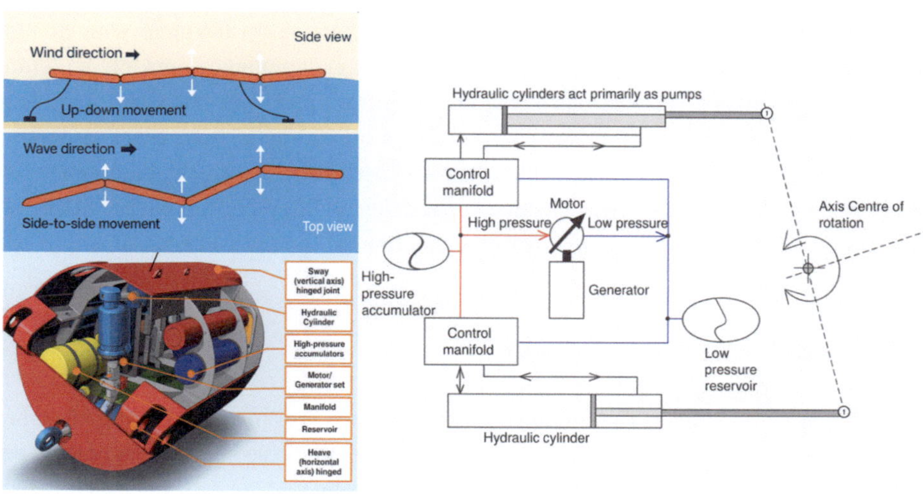


**Fig. 3.3** Illustration of an attenuator WEC, following the principle of the Pelamis WEC [2]

m long and 3.5 m in diameter, linked by hinged joints. As waves traveled along the length of the device, the motion at the joints drove hydraulic rams that pressurized fluid to spin hydraulic motors and generate electricity. A full-scale Pelamis device had a rated power capacity of 750 kW. In 2008, a three-device array was deployed off the coast of Agucadoura, Portugal, delivering a total of 2.25 MW. Although the project was eventually decommissioned and Pelamis Wave Power entered administration in 2014, the demonstration provided invaluable insights into offshore deployment, grid connection, and survivability.

Attenuator WECs offer several technical and operational advantages. Their elongated, flexible form allows them to distribute loads more evenly, potentially improving survivability during storms compared to point absorbers or fixed structures. By being aligned with the direction of wave propagation, attenuators can extract energy from long-crested swells common in offshore settings. The modular design consisting of multiple linked segments also facilitates scalability, ease of transportation, and simplified maintenance, as individual modules can be towed to shore for repairs. Moreover, their low profile and submerged orientation mean they have minimal visual impact, which is beneficial in regions sensitive to coastal aesthetics or tourism.

Despite these advantages, attenuator devices face several challenges. Their structural complexity and large physical footprint make manufacturing and deployment costly, especially when considering the robust materials needed to withstand saltwater corrosion, fatigue loading, and biofouling. Maintaining joint integrity and sealing hydraulic systems in a dynamic ocean environment is technically demanding. In addition, energy capture is optimal only when wave directions are closely aligned with the length of the device, making performance highly dependent on wave climate and potentially limiting their effectiveness in regions with multidirectional or short-period waves.

Looking ahead, attenuator WECs hold considerable promise, particularly for offshore renewable energy hubs. Their distributed design can potentially integrate with floating wind platforms or provide power for offshore aquaculture, communications infrastructure, or hydrogen production. The lessons learned from Pelamis and other trials have paved the way for second-generation attenuator designs that prioritize modularity, adaptive control, and advanced materials to enhance longevity and performance. As offshore renewable energy markets expand and costs decline through technology convergence, attenuators could become a valuable part of hybrid energy systems.

### 3.1.4 Oscillating Water Column

Oscillating Water Column (OWC) wave energy converters are among the most researched and deployed types of wave energy technology. These systems use the rise and fall of sea water within a partially enclosed chamber to compress and decompress air, which is then forced through a turbine connected to an electrical generator. As incoming waves push water into the chamber, air is expelled through the turbine; as the wave recedes, water falls and air is drawn back in, continuing the airflow in the reverse direction. Bidirectional turbines, such as the Wells turbine or impulse turbines with self-rectifying airflow, allow continuous power generation regardless of the air direction.

OWC devices can be classified into two primary types based on their placement: fixed (shoreline or nearshore) and floating (offshore). Figure 3.4 presents these two types, respectively. Fixed OWCs, such as those integrated into breakwaters or coastal cliffs, benefit from infrastructure access and ease of grid connection. Floating OWCs, on the other hand, offer deployment flexibility in deeper waters where wave energy is typically stronger and more consistent.

Floating OWCs, such as the Backward Bent Duct Buoy (BBDB), introduce additional design complexity but enable greater energy capture from offshore sites. The BBDB features an air chamber connected to a duct that is bent backward relative to incoming waves. This configuration creates a pressure differential as waves pass under the buoy, drawing air through the turbine in a controlled manner. The backward bend increases air compressibility and maintains a more stable airflow, improving the efficiency of the turbine. Floating OWCs like the BBDB are moored offshore and ride with the wave motion, which can reduce some of the harsh loads experienced by fixed structures.

One of the key strengths of OWCs is the lack of moving parts in direct contact with seawater, improving durability and simplifying maintenance. Since the turbine and generator are typically housed above the waterline, they are more accessible than submerged mechanical components. OWCs can also be easily integrated into multipurpose structures such as breakwaters, jetties, or even floating platforms for aquaculture or offshore wind. In particular, floating OWCs present an opportunity for scalable, modular wave energy farms in remote marine environments.

Nonetheless, several challenges limit OWC adoption. The energy conversion process—wave motion to airflow to mechanical rotation to electricity—involves multiple stages, each with associated losses. Turbines must function efficiently under rapidly fluctuating, bidirectional airflow, which can limit their responsiveness and overall performance. Floating OWCs face additional issues such as complex mooring requirements, pitch and heave instability, and susceptibility to harsh ocean conditions. For BBDBs, maintaining optimal duct orientation relative to the wave direction is critical and can complicate deployment in multidirectional wave fields.

Despite these hurdles, OWCs have demonstrated reliable operation and are among the few WEC types with grid-connected projects. The Mutriku Wave Power Plant in the Basque Country of Spain is a shoreline OWC facility with 16 air chambers using Wells turbines, producing a combined 296 kW of power since 2011. It has supplied the grid continuously for over a decade, offering valuable long-term operational data. Another pioneering system was the LIMPET (Land Installed Marine Powered Energy Transformer) on the Isle of Islay, Scotland, which delivered 500 kW of rated power until its decommissioning in 2018.

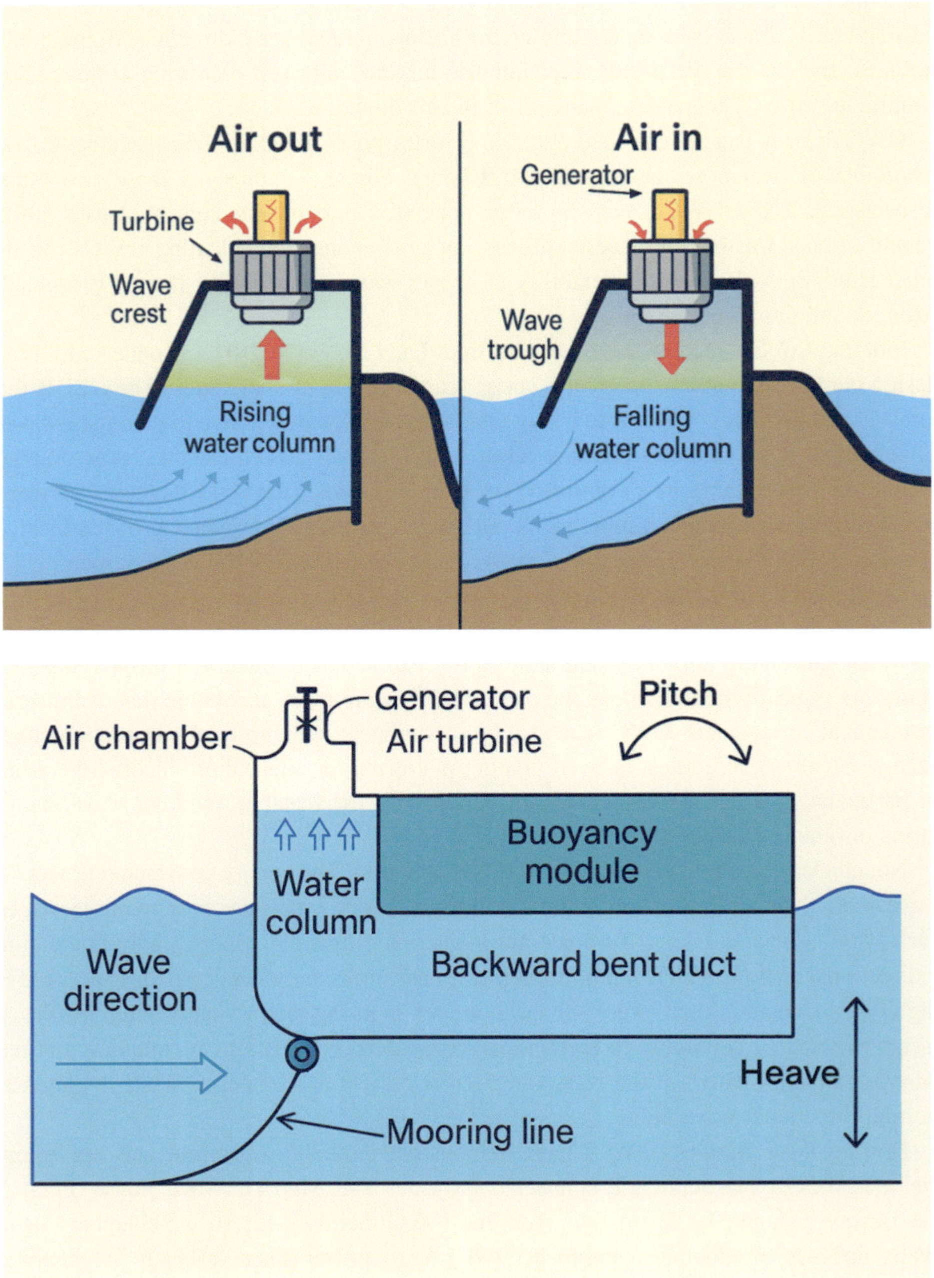


**Fig. 3.4** Illustration of fixed OWC WEC (upper) and floating OWC WEC (lower)

Floating OWCs, while still largely experimental, are gaining traction. The BBDB was originally developed in Japan and tested off the Japanese coast. Early prototypes demonstrated outputs in a few hundreds of kW range, depending on device scale and wave conditions. More recently, variations of BBDB have been explored by research teams in Korea and China, with experimental units deployed in scaled-down versions. Though no commercial-scale BBDB farms exist yet, they remain an active area of R&D, particularly for deployment in the Pacific and Indian Oceans where consistent offshore swell is available.

Looking to the future, floating OWCs like the BBDB could become critical for harnessing wave energy in deeper waters where fixed structures are impractical. They may also serve as modular units for remote island grids or contribute to hybrid marine energy platforms combining wind, wave, and solar technologies. As turbine technology evolves and mooring systems become more cost-effective, the commercial viability of floating OWCs is expected to improve. Innovations in adaptive control systems and directional tuning of ducts (for BBDBs) could further optimize energy capture.

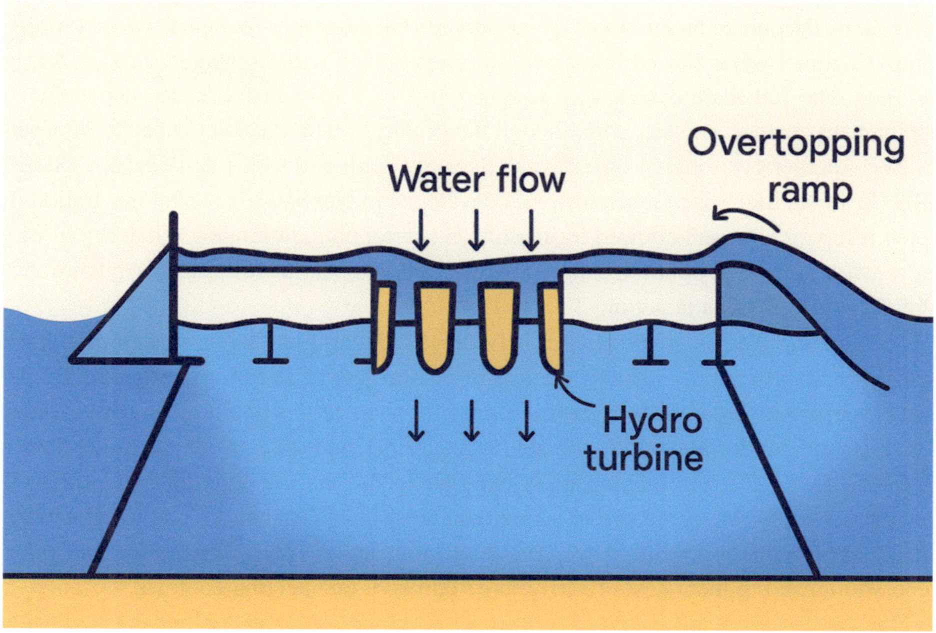


**Fig. 3.5** Illustration of an overtopping WEC

### 3.1.5 Overtopping

Overtopping wave energy converters operate on a fundamentally different principle than oscillating or attenuating devices. Instead of harnessing wave-induced motion directly, overtopping WECs capture wave energy by allowing incoming waves to spill over a ramp or structure into a raised reservoir. This impounded seawater—collected temporarily above the mean sea level—is then released back into the ocean through low-head hydro turbines, converting the gravitational potential energy of the elevated water into electricity. The concept is analogous to a small-scale hydroelectric dam operating in the marine environment, with waves acting as the water delivery mechanism, as shown in Fig. 3.5.

These devices can be deployed nearshore, integrated into breakwaters, or floated offshore. In fixed installations, a sloped or curved ramp channels waves into a basin. In floating configurations, the structure may use an inflatable or articulated lip to optimize overtopping in varying wave conditions. The efficiency of overtopping WECs depends on their ability to trap sufficient water volume at an adequate head height, which is influenced by wave steepness, device geometry, and site-specific hydrodynamics.

One of the most well-known overtopping devices is the Wave Dragon, a floating, slack-moored structure first tested in Denmark. The prototype, deployed off the coast of Nissum Bredning, featured two large curved reflectors that funneled waves toward a ramp. Captured water flowed into a central reservoir and was discharged through Kaplan turbines. The full-scale design was designed at 4 to 7 MW in full scale, depending on deployment location and sea state, though the prototype had a smaller capacity of around 20 kW. Another overtopping-type device, Société Hydro-Électrique de l'Océan's (SHEO) SSG (Seawave Slot-cone Generator), was developed in Norway for coastal integration. Its tiered reservoir structure allowed for multilevel overtopping, improving efficiency in varying wave conditions. A pilot installation with a 100 kW output was tested on breakwaters, with full-scale designs targetting 300–500 kW.

Overtopping WECs offer distinct advantages. First, they rely on well-established hydroelectric principles, enabling the use of proven low-head turbine technologies. The energy storage inherent in the reservoir allows for short-term smoothing of power output, which is valuable for integrating variable wave energy into the grid. Overtopping devices also tend to have fewer moving parts exposed to the marine environment, potentially reducing maintenance compared to submerged mechanical systems. When integrated into breakwaters or harbor walls, they serve dual purposes of coastal protection and power generation, making them cost-effective in certain civil engineering contexts.

However, overtopping systems also face significant design and deployment challenges. Their performance is highly sensitive to wave height and steepness, meaning they are best suited to energetic, long-period swells. In calmer or short-period seas, overtopping rates may be too low to justify installation. Floating versions must be carefully stabilized to maintain the reservoir above sea level, which adds structural and mooring complexity. Additionally, large surface footprints and visible profiles can be problematic in crowded or environmentally sensitive coastal zones. Turbulent flow during overtopping and release can also cause wear or biofouling in turbine passages, requiring regular maintenance.

Despite these limitations, overtopping WECs remain a promising technology—especially for multi-functional coastal infrastructure. The Wave Dragon project, while not commercialized, demonstrated that large-scale offshore overtopping is feasible, and the underlying technology has been proven in laboratory and pilot studies. Current research focuses on optimizing ramp geometry, improving mooring systems for floating versions, and integrating real-time control of reservoir levels to improve generation efficiency. There is also renewed interest in combining overtopping devices with floating breakwaters or offshore platforms for aquaculture, wind support, or desalination.

In the future, overtopping WECs could find a niche in hybrid marine energy systems or coastal adaptation projects where flood protection and renewable energy production are combined. As climate change increases pressure on coastal infrastructure and demand for resilient, low-carbon solutions rises, overtopping WECs may become part of a broader ecosystem of engineered shoreline systems that serve both protective and energy functions. While less versatile than some other WEC types, their ability to buffer short-term energy fluctuations and co-locate with existing marine structures makes them a potentially valuable component of a diversified ocean energy strategy.

### 3.1.6 Pressure-Based and Bulge Wave

Pressure-based and bulge wave energy converters utilize subsurface forces generated by passing ocean waves to create electricity. While both approaches rely on pressure fluctuations beneath the ocean surface, their internal mechanisms differ. This category includes traditional pressure differential WECs, such as the Archimedes Wave Swing, as well as bulge wave converters, like the Anaconda device, which generate internal waves within a flexible structure.

Pressure differential WECs typically consist of submerged structures—such as domes, membranes, or pistons—that respond to the alternating pressure field caused by passing wave crests and troughs. For instance, the Archimedes Wave Swing (AWS) features a large, submerged cylindrical chamber anchored to the seabed. As wave-induced pressure changes occur overhead, the upper floating portion of the device oscillates vertically, converting this motion into electrical energy via an internal generator. A 2 MW prototype was deployed in Portugal in the early 2000s, and a later 250 kW version was tested at EMEC in Scotland. These systems benefit from being largely hidden below the surface, minimizing visual impact and protecting key components from storm damage and surface corrosion.

Bulge wave converters, most notably represented by the Anaconda WEC, offer a different take on pressure-based energy capture. Anaconda consists of a long, water-filled rubber tube that floats just beneath the surface and is oriented parallel to the wave direction. As waves pass overhead, they generate pressure pulses at the front of the tube, which travel as bulge waves along its length. These internal waves grow in amplitude through resonance with the surface wave field and drive a power take-off system at the far end, as illustrated in Fig. 3.6. Developed by Checkmate SeaEnergy in collaboration with the University of Southampton, the Anaconda was envisioned at full scale to be about 200 m long and produce up to 1 MW of power. Although it remains in the prototype stage, it has attracted interest for its mechanical simplicity and potential low-cost deployment.

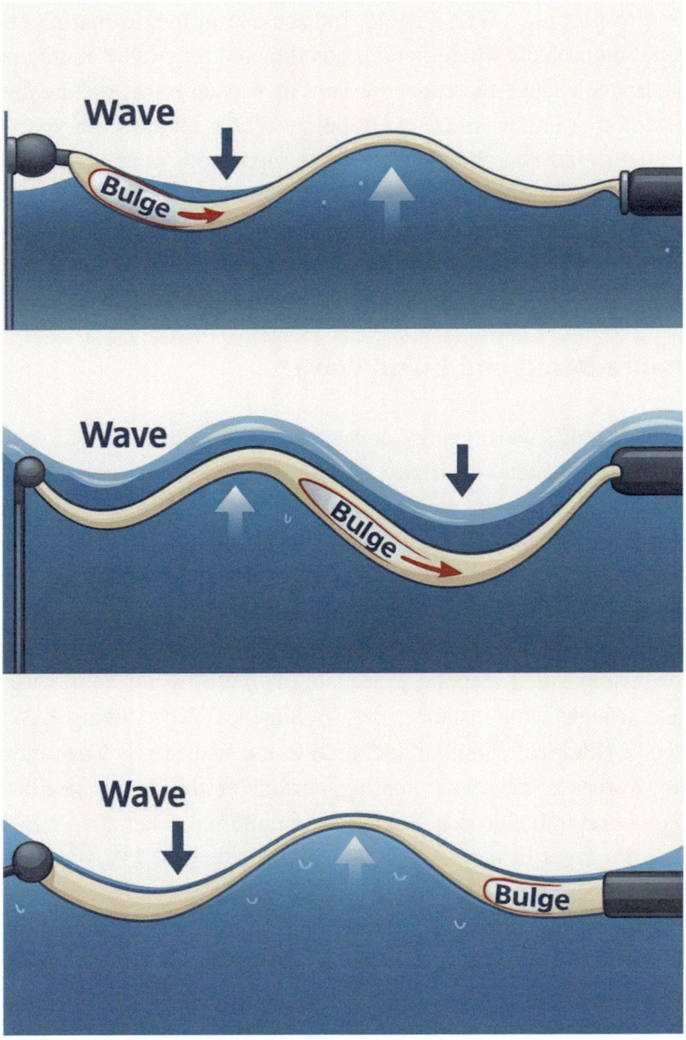


**Fig. 3.6** Illustration of a bulge WEC, following the principle of the Anaconda device

These technologies offer several shared benefits. Being submerged or semi-submerged, they are less exposed to storm loading and have a low visual profile—making them attractive for sensitive marine environments. They typically feature fewer moving parts than surface-based WECs, which could result in lower maintenance needs and improved reliability. Their modular nature allows for deployment in arrays, and their forms can be adapted to different wave climates and seabed types.

However, both approaches face technical and operational challenges. The underwater location complicates inspection and maintenance, often requiring ROVs or diver support. For pressure differential devices, the magnitude of usable pressure change may be insufficient in shallow or low-energy wave sites, limiting geographical applicability. In the case of Anaconda, the behavior of long, flexible tubes in real sea states introduces complex fluid–structure interaction problems. The materials used must strike a balance between flexibility, durability, and resistance to marine biofouling and UV degradation—making the choice of elastomers critical. Additionally, both technologies are still maturing, with limited long-term data from full-scale deployments.

Despite these challenges, the future potential of pressure-based and bulge wave WECs remains promising. They are well suited for offshore or deepwater environments where space is abundant and visual impact is a concern. Innovations in materials science, control systems, and hydrodynamic modeling are likely to improve their performance and resilience. There is also growing interest in integrating these systems with offshore infrastructure, such as aquaculture facilities or autonomous underwater observatories. While commercial-scale deployment may still be several years away, the distinct advantages of these designs—especially their mechanical simplicity and low-profile configuration—position them as valuable contributors to a diverse and adaptable wave energy sector.

### 3.1.7 Piezoelectric

Piezoelectric WECs represent an emerging class of technologies that utilize the piezoelectric effect, a phenomenon in which certain materials generate an electric charge when mechanically stressed, to convert the motion of ocean waves into electrical energy, as manifested in Fig. 3.7. Unlike traditional WECs that rely on large mechanical systems such as pistons, turbines, or hydraulic rams, piezoelectric devices offer the promise of compact, solid-state energy harvesting. This makes them particularly attractive for low-power marine applications and for deployment in challenging environments or remote locations where maintenance access is limited.

In a typical piezoelectric WEC, flexible or elastic structures embedded with piezoelectric materials, such as lead zirconate titanate (PZT), polyvinylidene fluoride (PVDF), or newer ceramic and polymer composites, are placed in the marine environment where they deform in response to wave motion or hydrodynamic pressure. As the materials

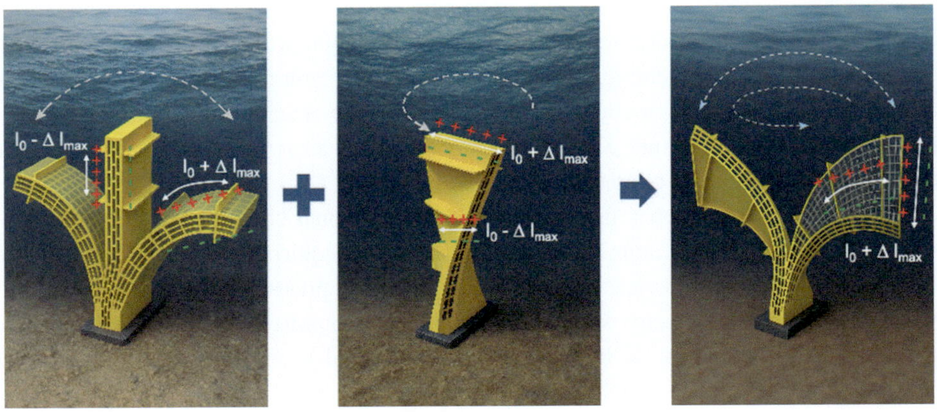


**Fig. 3.7** Illustration of the piezoelectric WEC, inspired by the NREL [3]

bend, twist, or compress under wave forces, they generate voltage across their surfaces, which can be harvested and conditioned for storage or immediate use. Designs range from small-scale energy harvesting strips attached to buoys, to more experimental structures like submerged flaps, tensioned cables, or flexible membranes anchored to the seafloor or suspended in the water column.

The key advantages of piezoelectric WECs lie in their mechanical simplicity, scalability, and low maintenance requirements. Because there are few moving parts, the systems are less prone to mechanical failure, making them well suited for long-term autonomous deployments. Additionally, piezoelectric devices can be fabricated at small scales, enabling their integration into sensor networks, underwater monitoring stations, or marine Internet-of-Things (IoT) platforms. They are especially useful in low-energy environments where larger, traditional WECs would be inefficient or uneconomical to install.

However, these systems also face notable limitations. The most significant challenge is their low power output—individual piezoelectric units typically generate electricity in the milliwatt to watt range, making them insufficient for large-scale grid integration without extensive scaling or aggregation. Energy conversion efficiency is also relatively low, and the electrical signals produced can be of high voltage but very low current, requiring specialized power electronics for conditioning and storage. Additionally, piezoelectric materials can be brittle, chemically sensitive, and expensive—especially ceramics that require complex fabrication or rare-earth elements. Durability in the marine environment remains a concern, particularly for polymer-based materials that may degrade under UV exposure or biofouling.

Despite these challenges, there has been growing research interest and early-stage prototyping around the world, primarily in the laboratory stage. The potential of piezoelectric WECs may lie not in large-scale electricity generation, but in niche, distributed, and hybrid energy systems. Advances in piezoelectric material science—such as the development of lead-free ceramics, flexible composites, and self-healing polymers—could enhance both performance and durability. Future deployment scenarios could involve arrays of piezoelectric strips integrated into marine structures like breakwaters, mooring lines, or floating platforms, passively generating electricity for lighting, sensors, or wireless communication systems. Their silent operation, low visual footprint, and ease of integration make them particularly suitable for environmental monitoring, defense, and marine research.

## 3.2 Wave Farm

Wave farms, arrays of WECs, represent a strategic approach to harvesting ocean energy by clustering multiple devices within a defined sea area. Much like offshore wind farms, the concept of a wave farm is not only about individual device performance, but also about the efficient use of valuable offshore space. By concentrating WECs into an organised array, developers can achieve economies of scale in permitting, monitoring, maintenance logistics, and grid connection, while minimising the overall footprint in marine environments. Careful layout design can also balance energy yield with navigation safety, ecological considerations, and compatibility with other maritime activities.

One of the strongest motivations for developing wave farms is their potential for co-location and shared infrastructure. WEC arrays can be sited alongside other offshore energy systems—such as wind farms, floating solar arrays, or tidal devices—allowing the sharing of transmission cables, mooring systems, and maintenance vessels. This multi-use approach reduces capital and operational costs while improving the utilisation of offshore grid connections. It also enables the creation of hybrid renewable energy hubs that make better use of ocean space, diversify energy output profiles, and increase the overall reliability of supply. Such synergies are particularly appealing in a world where both marine space and the capital for offshore projects are limited, making wave farms a logical step toward more integrated and sustainable ocean energy systems.

One of the defining and most technically complex aspects of wave farms is the wave farm effect. WECs interact with a common dynamic medium—water waves, which are themselves altered by the presence of each device. Ocean waves consist of alternating crests and troughs, and the power they carry is proportional to the square of the wave height. When a wave encounters a WEC, part of the wave's energy is extracted, and the rest is reflected, diffracted, or transmitted. These interactions modify the wave field around the device, affecting the wave characteristics that reach downstream WECs in the array. This creates a nonlinear feedback loop between energy extraction, wave propagation, and device performance—a phenomenon broadly referred to as the wave farm effect.

If an array can be configured so that the diffracted and radiated waves from one device align their crests with the incoming crests at the location of a downstream device, the resulting wave height at that device will be greater than in the unmodified incident wave field. Similarly, aligning troughs to coincide with troughs can amplify wave motion in a predictable way. This principle allows engineers to "focus" wave energy into specific points within the array, creating localised zones of higher energy density, as shown in Fig. 3.8. This effect doesn't exist for wind farms where interactive effects between turbines are mostly destructive.

Practical strategies to achieve constructive interactions within a wave farm begin with phased spacing. This involves positioning WECs at distances tuned to the dominant wave wavelength so that the scattered and radiated waves from one device arrive in phase with the incoming waves at another. When the timing is right, these overlapping wave fields combine constructively, boosting wave heights and energy at targeted points in the array. Another approach is directional steering, where devices are oriented or actively controlled to shape the direction of their radiated wave fields. By steering energy toward neighbouring units, the array can concentrate wave power in select locations, increasing the downstream devices' capture potential. This method is particularly effective in sites with consistent wave directionality, where predictable steering angles can be maintained. Hybrid configurations offer a way to broaden the range of conditions over which constructive interference can occur. By combining devices with different hydrodynamic responses—such as point absorbers, oscillating water columns, and attenuators—an array

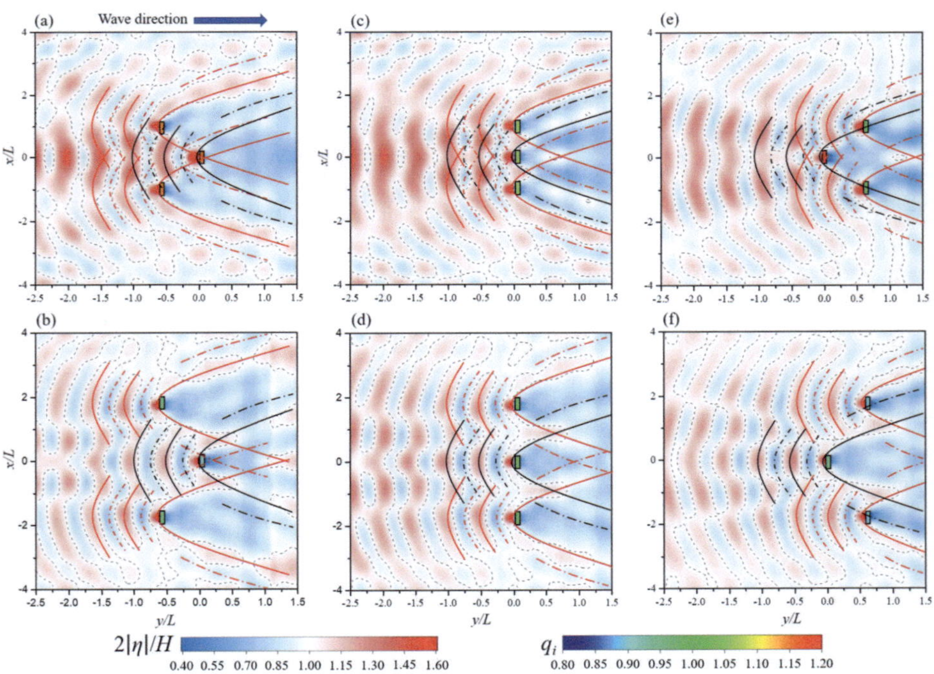


**Fig. 3.8** Demonstration of where waves are amplified (Red) or weakened (Blue) in a wave farm [4]

can exploit a wider band of the wave spectrum. This helps maintain beneficial interference effects even as sea states change. Sequential energy buildup can be achieved by arranging devices so that each successive row benefits from the wave focusing effects of those upstream. In this configuration, energy is gradually concentrated through the array, with each stage incrementally increasing the available power for the next. When optimised, this creates a cascading amplification effect that can significantly raise the total energy yield of the farm.

Several strategies have been proposed to achieve optimal layouts. These include staggered or offset configurations, non-uniform spacing, and adaptive positioning that responds to seasonal wave climate variations. Computational tools such as boundary element models (BEM), computational fluid dynamics (CFD), and spectral wave models like SWAN and WAM are often used to simulate these interactions and optimize deployment schemes. Recent research has also considered hybrid layouts, where different types of WECs are deployed together to balance energy capture with environmental impact and structural resilience. Moreover, directionality of waves plays a critical role: in environments with a dominant wave direction, linear rows of WECs may be beneficial, while omnidirectional sites might favor radial or circular arrays.

However, achieving these benefits is complicated by the inherently irregular nature of ocean waves. Real seas are composed of many wave components with different wavelengths, heights, and directions, and these characteristics can change on timescales from seconds to seasons. A layout optimised for one dominant wavelength may not achieve perfect constructive interference when the sea state shifts, and in some conditions, partial destructive interference may occur. For this reason, array design must focus not on maximising performance at a single idealised condition, but on achieving consistently improved energy efficiency over the long term. This requires detailed spectral wave modelling, probabilistic performance assessment, and in some cases adaptive control strategies to ensure that, across the full range of operating conditions, the farm delivers a meaningful net gain in output compared to a non-optimised layout.

The need to consider wave farms rather than individual devices is not only technical but also economic and regulatory. A single WEC installation may struggle to justify the substantial costs and administrative effort associated with permitting, grid connection, and environmental monitoring. By contrast, deploying devices as part of a coordinated farm enables economies of scale, spreads fixed costs over multiple units, and increases project bankability, making the investment case more compelling for both private and public stakeholders. From a planning perspective, wave farms also allow for more responsible and efficient use of marine space, consolidating energy generation within a defined footprint rather than scattering isolated devices over a wider area. This spatial concentration can reduce navigational risks, simplify exclusion zone management, and help avoid sensitive ecological zones through careful site selection and layout optimisation.

Well-planned wave farms can also minimise environmental impacts compared to ad-hoc, standalone deployments. Grouping devices allows for a single, shared export cable to shore, reducing the total length of subsea cabling and the associated disturbance to benthic habitats. Shared infrastructure—such as moorings, monitoring systems, and maintenance vessels—further reduces the overall environmental footprint of the project. Successful precedents include the Wave Hub project in Cornwall, UK, which, though still in its demonstration phase, provides subsea grid-connection and berth infrastructure for multi-device wave-energy arrays. The Mutriku breakwater wave plant in Spain, which integrates 16 OWC chambers into an existing coastal structure, producing 296 kW in total while avoiding new offshore foundations. More recently, developers such as CorPower Ocean, Ocean Energy, and Eco Wave Power are designing their technologies from the outset with farm-scale deployment and co-location (wave + other renewables) in mind, enabling them to align with marine spatial planning frameworks, support hybrid offshore renewable hubs, and deliver higher overall energy yield while respecting the surrounding marine environment.

WECs and offshore wind turbines share certain environmental considerations, yet their impacts differ in both nature and scale due to the way each technology interacts with the marine environment. Offshore wind farms typically involve tall fixed-bottom or floating structures concentrated in discrete points, creating a strong visual presence and requiring substantial foundations or anchors. Installation of fixed-bottom wind turbines often involves pile driving, which generates intense impulsive underwater noise capable of disturbing or displacing marine mammals and fish over large distances. WECs, by contrast, are generally lower in profile, often semi-submerged or fully submerged, and have minimal visual impact. Many WEC designs are anchored or ballasted without pile driving, which reduces installation noise and sediment disturbance. However, their arrays may have more numerous seabed contact points spread over a wider area.

WECs extract energy directly from waves, thereby altering local wave height, period, and direction in their lee. Large arrays can change patterns of sediment transport and coastal processes, potentially influencing erosion or habitat conditions near shore—effects that offshore wind turbines generally do not produce to the same extent. On the other hand, both technologies can create artificial reef habitats by introducing hard surfaces that attract sessile marine organisms and fish. Offshore wind poses aerial hazards for birds and bats due to spinning blades, while WECs eliminate this risk but may introduce underwater collision or entanglement hazards for marine mammals, particularly with moving components or mooring lines. In short, WECs tend to have lower visual impact, reduced bird strike risk, and less installation noise compared to offshore wind, but they require careful consideration of seawater movement, seabed footprint distribution, and potential underwater entanglement risks.

## 3.3 WECs Around the World

Europe remains the global leader in wave energy development, with around 13 MW of cumulative installed capacity an estimated currently operational capacity of around 1 MW and a large portfolio of devices tested at sea over the last two decades. Countries such as the United Kingdom, Portugal, Spain, Denmark, and Sweden have maintained a proactive role in supporting early-stage WEC technologies through public funding schemes, EU-backed research programmes (e.g., Horizon Europe), dedicated test sites such as EMEC in Orkney and BiMEP in Spain, and grid-connected demonstration projects. Notable long-running installations include the Mutriku breakwater plant in Spain, which integrates 16 oscillating water column (OWC) chambers into a harbour structure to produce a combined 296 kW, and has operated continuously since 2011, serving as a valuable dataset for long-term OWC performance. Portugal's Aguçadoura wave farm, though only briefly operational in 2008 with 2.25 MW of Pelamis terminator-type WECs, remains a landmark in the sector's history as the world's first grid-connected wave farm. Denmark's Wave Dragon and Sweden's Seabased point absorber arrays further underscore Europe's early leadership and the diversity of its technological approaches. More recent deployments include CorPower Ocean's HiWave-5 project in Portugal, which entered operational testing in 2023, and AW-Energy's WaveRoller in Peniche, Portugal, which has provided grid-connected output and contributed to commercial readiness validation.

In Asia, installed capacity remains modest, but the region has demonstrated distinctive technological innovations and strong research capabilities. Japan has long been active in wave energy, pioneering floating oscillating water column devices such as the Backward Bent Duct Buoy (BBDB) and continuing to test new designs through the New Energy and Industrial Technology Development Organization (NEDO). Japanese researchers have also explored hybrid offshore platforms combining wind, solar, and wave modules. China, while was primarily investing in offshore wind, has increased wave energy research through state-backed pilot projects, coastal university laboratories, and demonstration sites in provinces such as Shandong and Guangdong. The Dalian University of Technology developed point absorber and several nearshore oscillating flap devices that have undergone sea trials, though commercial-scale deployments remain pending. In South Korea has also initiated small-scale demonstrations, often as part of multi-use coastal protection structures.

Oceania, particularly Australia, has played host to high-profile demonstration projects, most notably Carnegie Clean Energy's CETO system. This subsurface, pressure-based point absorber has been tested at Garden Island in Western Australia, including integration with a wave-powered desalination plant. While no project has yet achieved sustained commercial operation, the CETO trials, along with Bombora Wave Power's mWave membrane-based device currently undergoing commissioning in Wales before planned Australian deployment, highlight the region's role in advancing novel WEC architectures. Government-backed research funding and the Australian Renewable Energy Agency

(ARENA) continue to support development, and total operational capacity, though still low, is bolstered by strong industry–research partnerships.

In North America, wave energy remains in the R&D and pre-commercial testing phases, but infrastructure is improving. The United States is home to the PacWave South test facility in Oregon, a fully permitted and grid-connected site expected to host multiple WEC types from 2025 onwards. Ocean Power Technologies (OPT) continues to operate and refine its PB3 PowerBuoy for both renewable energy and maritime surveillance applications. The U.S. Department of Energy's Water Power Technologies Office and the National Renewable Energy Laboratory (NREL) have expanded modelling and testing capabilities to accelerate commercial readiness. In Canada, activity is focused on the Pacific coast, where the West Coast Wave Initiative has conducted detailed resource assessments, and several universities are exploring device concepts tailored to local wave climates.

South America and Africa are at an early stage of wave energy exploration, but resource potential is high. In Chile, feasibility studies by universities have confirmed strong wave climates along the southern Pacific coast. South Africa has conducted coastal wave energy assessments and small-scale mechanical prototype tests. While neither region yet hosts commercial or grid-connected wave farms, increasing involvement in international collaborative projects suggests that pilot deployments could materialise in the next few years.

## 3.4 Key Challenges

Despite their significant potential, WECs face a broad set of technical, economic, and environmental challenges that have so far hindered their large-scale commercial adoption. These obstacles span the full lifecycle of WEC systems—from design and deployment to operation and decommissioning—and are fundamentally tied to the harsh and dynamic nature of the marine environment in which they operate.

### 3.4.1 Energy Efficiency and Structural Integrity

One of the most persistent challenges lies in energy efficiency and variability. Wave energy is inherently complex and irregular. The spectral composition of ocean waves varies with wind patterns, weather systems, and geography, resulting in highly unpredictable sea states. WECs must therefore be designed not just to extract energy from a single type of wave but to perform consistently across a wide range of frequencies, heights, and directions. This often leads to compromises in performance. Many WECs achieve relatively low energy conversion efficiency, especially when operated outside their optimal wave conditions. Moreover, capturing energy efficiently from both the heave (vertical), surge (horizontal), and pitch (rotational) motions of waves remains a technical challenge that few systems have fully solved.

To address this problem, new approaches are focusing on adaptive control systems capable of automatically tuning WECs in response to real-time wave conditions. By adjusting key mechanical or hydraulic properties—such as the stiffness of mooring lines, the damping of power take-off systems, or the geometry of moving components—these systems can optimise the device's natural frequency to align with the dominant wave period at any given time. Such tuning can significantly improve energy capture across a wider range of sea states without sacrificing survivability in extreme conditions. In some designs, the control system can also switch operating modes entirely during storms, prioritising structural protection over power production. This dynamic adaptability not only boosts overall energy yield but also reduces wear and tear, potentially lowering maintenance costs and extending the operational lifetime of the device.

Structural durability is another major hurdle for wave energy converters. To harness more energy, devices must operate in environments with larger, more powerful waves—yet the very conditions that offer the highest energy potential also impose the greatest structural loads. In these high-energy environments, devices are subjected not only to constant cyclic mechanical forces, sometimes numbering in the millions of load cycles per year, but also to occasional extreme events that can exceed the design envelope many times over. Rogue waves, storm surges, and long-period swells can combine to produce sudden, concentrated forces capable of damaging hulls, breaking moorings, or fatiguing critical components. This creates an inherent tension: the most lucrative wave climates are also the harshest for long-term survivability.

Mitigating this challenge requires both innovative device architectures and intelligent material choices. Designs such as oscillating water columns (OWCs), pressure-based devices, and bulge wave systems have the advantage of decoupling much of the structural loading from the energy conversion process—for instance, by transferring wave motion into enclosed air or water chambers rather than directly into mechanical structures. This reduces the magnitude of instantaneous loads on the primary structure, lowering fatigue stresses and improving reliability. Similarly, piezoelectric systems, while still in early-stage development, hold promise for their ability to convert wave-induced strains directly into electrical energy using solid-state materials, avoiding many of the moving parts and large structural members that are prone to mechanical wear.

Another avenue for load mitigation lies in the use of soft or flexible structural elements that can deform under extreme conditions, absorbing energy rather than resisting it with rigid strength. This could include compliant membranes, elastomeric joints, or modular floating segments that naturally adapt to the wave field, spreading loads more evenly and avoiding catastrophic failure modes. The combination of load-decoupling energy conversion principles (like OWCs) with strategically flexible structural elements represents one of the most promising pathways toward truly durable, economically viable wave energy systems in large-wave environments.

Coupled closely with structural challenges are the maintenance and accessibility issues unique to marine energy. Most WECs are deployed far offshore, often in high-energy wave environments where safe weather windows for vessel operations are limited. This makes routine inspections, repairs, and part replacements logistically complex, time-consuming, and costly. Unlike offshore wind turbines—where a significant portion of equipment is above the waterline and can be serviced using drones, rope access teams, or small crew transfer vessels—WECs typically have most of their critical components submerged. This means that maintenance often requires dive teams or remotely operated vehicles (ROVs), which are slower, more expensive, and more climate-dependent than above-water interventions.

Towing devices to shore for servicing is rarely feasible for large or bottom-fixed systems, and in-situ work at sea demands specialised ships, heavy-lift capabilities, and experienced marine crews. Underwater servicing also brings visibility challenges, strong currents, and biofouling that can obscure inspection points or hinder access to components. In the future, robotics and autonomous systems may play a pivotal role in addressing these issues. ROVs and autonomous underwater vehicles (AUVs) equipped with high-resolution imaging, cleaning tools, and manipulators could conduct regular inspections, perform minor repairs, and even swap out modular components without the need for large service vessels. There is also potential for semi-autonomous "docking stations" on or near WEC arrays, where maintenance robots could recharge, upload inspection data, and be deployed on demand. While these concepts are still emerging, their adoption could significantly reduce downtime, improve safety by limiting human exposure to hazardous offshore conditions, and lower lifetime O&M costs for wave energy projects.

### 3.4.2 Costs

One of the most persistent challenges facing wave energy development lies in achieving economic viability. Capital expenditure (CAPEX) for pre-commercial WECs typically can reach several thousands of USD per kW of installed capacity, depending on the deployment stage—first arrays are at the higher end, while second arrays and commercial-scale projects approach the lower range. Operational expenditure (OPEX) remains substantial, around tens to hundreds of USD per kW per year, again varying by deployment maturity [5]. These figures underline the economic headwind faced by wave energy, compounded by higher investor risk premiums and limited access to commercial financing.

Not all WEC technologies perform equally in terms of cost efficiency, and the differences in levelized cost of energy (LCOE) can be substantial. A techno-economic review of leading designs highlights this variation: the Wave Dragon overtopping device achieves an estimated LCOE of around $355/MWh, while the Pelamis terminator-type system comes in significantly higher at approximately $800/MWh. At the other extreme, the AquaBuOY point absorber has reported LCOE values approaching \$3200/MWh, reflecting the high costs and relatively low energy yields of small-scale point absorber technologies in early development stages [5]. Oscillating water columns (OWCs), particularly those integrated into fixed structures such as breakwaters, tend to achieve more favourable economics due to lower installation and maintenance costs; published values for nearshore OWCs are often in the \$500/MWh region, with Mutriku's breakwater OWC plant serving as a long-running example of cost stability over more than a decade of operation [6]. Scale and financing terms have an outsized influence on these cost figures. Modelling studies suggest that with parameter optimisation—covering aspects such as advanced control strategies, constructive array configurations, and favourable financing—LCOE values could drop substantially.

Overall, while many WEC designs remain more expensive than mature renewables such as offshore wind, the combination of optimal site selection, scale-up, targeted control strategies, and hybrid infrastructure integration offers a pathway for certain designs to approach commercially viable LCOE levels within the next decade.

### 3.4.3 Power Take-Off

PTO systems lie at the heart of WECs, yet they remain one of the main bottlenecks to commercial viability. Their function—transforming the slow, irregular, and often unpredictable motion of waves into steady electrical output—is technically demanding in a marine environment where corrosion, biofouling, and extreme loads are unavoidable. The central challenge is achieving high efficiency, reliability, and maintainability under these conditions without making the system prohibitively expensive or complex. Many existing PTO designs face trade-offs between mechanical robustness, energy capture efficiency, and the ability to survive storm events.

Hydraulic PTOs, which pressurise fluid to drive a motor-generator set, offer smooth and controllable power output and can incorporate energy storage through hydraulic accumulators. However, they introduce multiple conversion stages, each with efficiency losses. More critically, they rely on extensive sealing to prevent leakage—an ongoing maintenance burden in saltwater environments. Component wear, susceptibility to contamination, and the need for regular servicing make long-term deployment costly and operationally challenging.

Direct-drive PTOs convert the motion of a wave energy converter directly into electricity by attaching the moving part of the device straight to an electrical generator. In this context, the generator is a machine that turns mechanical movement into electrical power using electromagnetic induction, essentially, moving a set of magnets past coils of wire to create an electric current. For wave energy, this generator can be a linear generator, where the magnets and coils slide back and forth in a straight line (matching the up-and-down motion of waves), or a rotary generator, where the motion is converted into rotation before generating electricity. Because direct-drive systems do not use gearboxes, turbines, or hydraulic pumps, they have fewer moving parts and lose less energy in conversion. However, wave motion is naturally slow but forceful, so the generator must be designed to produce useful amounts of electricity at very low speeds. This usually means using large, powerful (and expensive) permanent magnets, which add cost and weight. The electricity that comes out is also irregular in frequency and voltage, so it needs advanced power electronics to make it stable and suitable for the grid.

Air-turbine PTOs, used in oscillating water column devices, sidestep many marine sealing issues by housing mechanical components above the waterline. Yet, they face aerodynamic inefficiencies, especially with bidirectional turbines like the Wells turbine, and their performance can drop sharply in low-energy seas. These characteristics limit their adaptability to diverse wave climates.

Piezoelectric PTOs, which convert mechanical strain directly into electrical energy using materials that generate voltage under deformation, represent a lightweight, solid-state alternative. Their corrosion resistance and lack of moving parts are appealing for marine use, particularly in small-scale or autonomous devices. However, their power output per unit volume is currently low, requiring large arrays or mechanical amplification systems to be viable for grid-scale WECs. Furthermore, piezoelectric materials can be brittle, raising durability concerns under repeated high-strain loading.

Looking forward, the main PTO challenge is not simply to improve efficiency, but to develop systems that maintain that efficiency across a wide range of sea states while surviving decades at sea with minimal maintenance. Emerging ideas such as variable-damping PTOs with real-time control, hybrid PTO architectures combining direct-drive with secondary energy storage, and advanced materials—including flexible piezoelectric composites—could help bridge the gap. Still, cost-effective reliability remains the defining hurdle, and solving it will be key to making wave energy commercially competitive.

## References

1. Qiao D., Haider R., Yan J., Ning D., and Li B. Wave Energy Converter. Encyclopedia (MDPI) (2020).
2. Henderson R. Design, simulation and testing of a novel hydraulic power take-off system for the Pelamis Wave Energy Converter. Renew. Energy 31, 271-283 (2006).
3. Mendoza N., Boren B. & Niffenegger J. Distributed Embedded Energy Converter Technologies for Marine Renewable Energy: A Technical Report. National Renewable Energy Laboratory, NREL/TP-5700-85158 (Aug. 2023).
4. Wang C., Wei Y., Chen W. & Huang L. Interactive effects of deformable wave energy converters operating in tandem in waves. Energy 320, 132905 (2024).
5. Guo C., Sheng W., De Silva D. G., and Aggidis G. "A Review of the Levelized Cost of Wave Energy Based on a Techno-Economic Model." Energies 16, 2144 (2023).
6. Peña A., Bidaguren I., Izquierdo U., Esteban G.A., Blanco J.M. & Albaina I. The Mutriku Breakwater Wave Plant: Improvements and Their Influence on the Levelized Cost of Electricity (LCoE). J. Ocean Eng. Technol. 39(2), 205–211 (2025).

# 4 Tidal Energy Converters

Tidal energy converters (TECs) harness one of the most predictable renewable energy sources on the planet—the movement of water driven by the gravitational pull of the Moon and Sun. Unlike wind and wave resources, which can fluctuate unpredictably from hour to hour, tidal currents follow regular astronomical cycles, making their timing and strength forecastable in advance. This predictability gives tidal energy a unique advantage in grid planning, allowing it to complement the variability of offshore wind farms and WECs.

Most modern TECs operate on principles similar to underwater turbines. Fast-moving tidal currents pass over rotor blades or hydrofoils, creating lift and drag forces that produce rotational motions. This mechanical motion drives an electrical generator, either directly or via a gearbox—converting the kinetic energy of the moving water into electricity. Because water is about 800 times denser than air, tidal turbines have the potential to generate significant power at much lower flow speeds than wind turbines, enabling more compact designs for equivalent output. However, the same density also means far greater structural loading, requiring robust engineering, corrosion-resistant materials, and designs that can endure decades of immersion in high-flow marine environments.

In contrast to wave energy, which captures the irregular oscillations of the ocean surface, tidal currents provide a relatively steady, unidirectional flow for hours at a time before reversing. Many TECs are built to operate bidirectionally or to yaw with the changing current direction. This regularity simplifies some aspects of control and power conversion, though installation and maintenance in high-flow tidal channels remain technically demanding and costly. Depth and seabed conditions also restrict site selection—devices must be placed in areas with both strong tidal currents and suitable foundation conditions, which are geographically limited compared to offshore wind.



L. Huang, *Ocean Renewable Energy Systems*, Synthesis Lectures on Renewable Energy Technologies, https://doi.org/10.1007/978-3-032-10576-9_4

Ocean-based TECs are usually deployed in coastal or estuarine areas where tidal flow is naturally concentrated, such as between islands, around headlands, or through narrow straits. These sites must be carefully chosen to balance energy yield, environmental impact, and navigational safety. Once installed, electricity is transmitted through subsea cables to onshore substations for conditioning and grid integration. While submerged operation gives TECs a low visual profile, making them less intrusive than offshore wind, they must be designed to minimise effects on marine ecosystems, sediment movement, and shipping routes.

As the technology matures, advances in materials, floating and modular platforms, and automated maintenance systems are improving durability and reducing costs. Novel designs, from oscillating hydrofoils to venturi-accelerated flow channels, are expanding the range of viable deployment sites. If these developments continue, tidal energy could take its place alongside wind and wave as a core component of a diversified offshore renewable energy mix, delivering reliable, predictable power that other sources cannot match.

## 4.1 Operating Principles

There are several main types of ocean-based tidal energy converters, which can broadly be grouped into two categories: kinetic energy systems and potential energy systems. Kinetic systems harness the motion of tidal currents, and within this category the most common designs are turbine-based and non-turbine devices. Turbine-based systems include horizontal-axis turbines, which resemble offshore wind turbines with blades rotating around a horizontal shaft aligned with the tidal flow, and vertical-axis turbines, which rotate around a vertical shaft and can accept flow from multiple directions, offering greater flexibility in bidirectional tidal environments, and some more special type of turbines such as Archimedes screws. Non-turbine kinetic devices capture energy without a rotating rotor; examples include oscillating hydrofoils, which use wing-like structures moving up and down in the flow to drive mechanical linkages and generate electricity, as well as kites and paddle wheels. Non-turbine designs are promising for their mechanical simplicity, potentially lower environmental impact, and suitability for shallow or low-flow environments. Each of these kinetic systems typically comprises components such as rotors or foils, a power take-off mechanism (usually a generator), a support structure (such as a monopile, floating platform, or gravity base), and a subsea cable that transmits the generated electricity to shore.

In contrast, potential energy systems exploit the difference in water levels between high and low tides (known as tidal range) to generate electricity. This group includes technologies such as tidal barrages, tidal lagoons. These systems work by impounding water during high tide and releasing it during low tide through turbines, converting the gravitational potential energy of the water mass into electrical energy. While more infrastructure-intensive, potential energy systems offer predictable, dispatchable power and have been proven at scale in several locations around the world.

### 4.1.1 Horizontal-Axis

Horizontal-axis tidal energy converters (HATs), as illustrated in Fig. 4.1, are one of the most widely developed and studied types of tidal energy systems used in ocean environments. These devices operate using the same fundamental principles as wind turbines, except they are placed underwater and harness the kinetic energy of tidal currents rather than air flow. The primary component of a HAT is a set of rotor blades mounted on a horizontal shaft, which is aligned with the direction of the tidal current. As water flows over the blades, it causes them to rotate, which in turn drives a generator either directly or via a gearbox to produce electricity. These systems are typically mounted on a fixed support structure such as a monopile, gravity base, or tripod on the seabed, although floating variants with mooring systems are also being explored.

The appeal of horizontal-axis tidal turbines lies in their relatively mature technology and proven energy conversion performance. They are efficient in unidirectional or bidirectional flow environments and can be designed to yaw or rotate to face the oncoming current during tidal reversals. One of the key advantages of HATs is their scalability—individual units can be deployed in arrays, much like wind farms, allowing developers to aggregate power production in locations with high tidal energy potential. Furthermore, their submerged installation reduces visual intrusion and allows for energy harvesting without occupying valuable land space.

**Fig. 4.1** Illustration of the Horizontal-axis tidal energy converters

A notable variation of horizontal-axis turbines is the ducted flow system, which incorporates a shroud or duct around the rotor to accelerate water flow through the blades. This design can increase energy capture, improve efficiency at lower current speeds, and offer additional protection to the rotor from debris or marine growth. Ducted turbines are especially attractive for sites where flow velocities are moderate or variable, as they can generate comparable power to larger open-rotor systems while reducing rotor diameter and structural loads.

However, HATs also come with several limitations. The underwater environment is mechanically and chemically harsh, leading to challenges related to corrosion, marine growth, and maintenance accessibility. Installation and retrieval often require specialized vessels and subsea infrastructure, which can raise capital and operational costs. Environmental concerns have also been raised, particularly regarding the potential for rotor blade collisions with marine fauna and the alteration of local hydrodynamics. Despite these concerns, ongoing research is focused on minimizing ecological impact through improved rotor designs, slower blade tip speeds, and environmental monitoring technologies.

Several significant projects around the world showcase the potential of horizontal-axis tidal energy. One of the most prominent is the MeyGen project in Scotland's Pentland Firth, which represents the world's largest planned tidal energy array. Phase 1A of MeyGen deployed four HATs with 4 turbines of 1.5 MW each, a combined capacity of 6 MW, and the project is eventually planned to scale up to nearly 400 MW. Another example is Nova Innovation's tidal array in Shetland, which operates several turbines rated at about 100 kW each. In Canada, the FORCE (Fundy Ocean Research Center for Energy) site in the Bay of Fundy serves as a test bed for HAT and other tidal technologies.

The future potential of horizontal-axis tidal energy converters is substantial, particularly as countries seek to diversify their renewable energy portfolios with reliable, predictable sources. As turbine designs become more cost-effective, robust, and environmentally sensitive, HATs are poised to play a significant role in coastal energy systems. Innovations in materials, remote maintenance, and modular construction are expected to drive down costs and increase deployment flexibility.

### 4.1.2 Vertical-Axis

Vertical-axis tidal energy converters (VATs), as illustrated in Fig. 4.2, are a class of tidal turbines designed with a rotor shaft oriented perpendicular to the direction of water flow and to the seabed. Unlike horizontal-axis turbines, VATs do not need to align themselves with the direction of the tidal current, making them particularly well-suited for environments with highly variable or multidirectional flows. The most common designs include Darrieus-type turbines, which have curved blades forming a helix or an "eggbeater" shape,

**Fig. 4.2** Illustration of vertical-axis tidal energy converters

and Savonius-type turbines, which use scooped blades to harness drag forces. These systems can be deployed either on fixed structures mounted to the seabed or on floating platforms depending on site conditions and design preferences.

One of the main advantages of vertical-axis tidal turbines is their ability to capture energy from tidal currents regardless of direction, which reduces the need for active yaw mechanisms and allows for operation in complex tidal environments such as estuaries or regions with eddies and cross-currents. Additionally, many VAT designs have a lower profile and can be placed closer to the seabed, minimizing navigation risks and visual impact. Their mechanical simplicity—particularly the fixed-axis orientation—can also translate to easier maintenance and reduced moving parts compared to horizontal-axis systems. Furthermore, VATs can have smaller dimensions than HATs, which may allow deployment in a wider range of sites, albeit with lower power density.

However, VATs typically suffer from lower energy conversion efficiency than horizontal-axis turbines, especially in high-speed unidirectional flows. The start-up torque, i.e., the ability of the turbine to begin rotating on its own, can be a challenge, often requiring auxiliary mechanisms or hybrid configurations. The asymmetric loading of the blades during rotation can lead to fatigue and increased wear, which complicates long-term reliability and structural design. Moreover, fewer VAT designs have reached commercial maturity, and large-scale demonstrations remain relatively limited compared to the horizontal-axis counterparts.

Several notable projects around the world have demonstrated vertical-axis tidal technologies at pilot scales. In France, HydroQuest's OceanQuest project tested a 1 MW vertical-axis tidal turbine at the Paimpol-Bréhat test site. HydroQuest also announced the FloWatt project that plans to test a 17 MW array of six VATs. Verdant Power in the United States has also experimented with vertical-axis designs in the East River in New York City, although its commercial focus later shifted toward horizontal-axis models due to efficiency considerations. In the UK, Achelous Energy is developing floating vertical-axis tidal turbines designed for deployment in near the water surface, leveraging modular VAT units and a low-footprint mounting system to enable scalable energy extraction.

Looking ahead, the future potential of vertical-axis tidal energy converters is closely linked to advances in design optimization, materials engineering, and computational modeling. Innovations in blade geometry, structural resilience, and hybrid control systems could help VATs overcome their traditional limitations and become competitive with more established technologies. Their adaptability to complex hydrodynamic environments makes them a promising option for specific tidal sites that would otherwise be unsuitable for horizontal-axis systems. As the global demand for diverse and site-specific renewable solutions grows, VATs may find increasing relevance in remote coastal communities, island nations, and low-depth tidal basins—contributing to a more resilient and diversified tidal energy sector.

### 4.1.3 Archimedes Screws

Archimedes Screw Tidal Energy Converters (ASTEC) are a type of kinetic tidal energy system that harnesses the flow of water through a helical screw to generate electricity. Unlike rotary turbines that rely on high-speed blades, ASTECs use a large, slowly rotating screw housed within a cylindrical casing. As tidal currents pass through the screw, the water's kinetic energy causes the helical blades to turn, driving a gearbox and generator to produce electrical power. The principle is conceptually similar to traditional Archimedes screws used for pumping water, but in reverse: instead of lifting water, the flowing water rotates the screw to generate electricity. These systems can be installed on the seabed or in shallow channels, making them particularly suitable for estuarine sites or locations with moderate tidal flows (Fig. 4.3).

One of the key advantages of Archimedes screw tidal devices is their mechanical simplicity and robustness. The slow rotation of the screw reduces mechanical wear, and the system can operate efficiently under low to moderate flow speeds (typically 1–3 m/s),

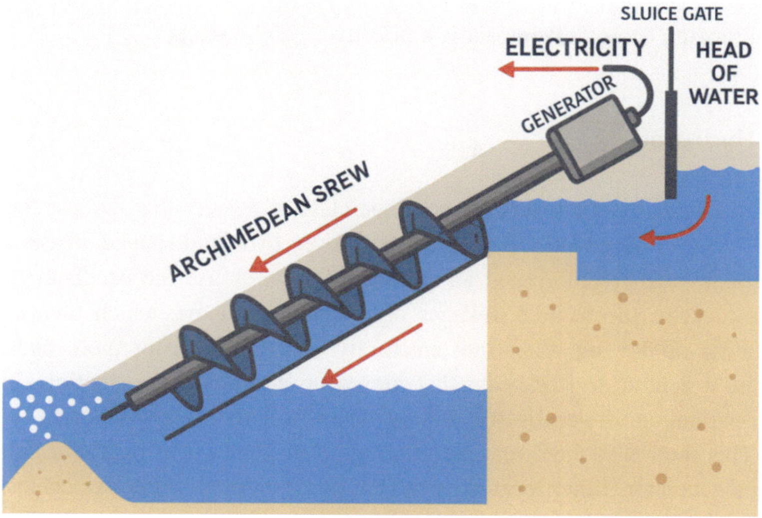


**Fig. 4.3** Illustration of an Archimedes screws tidal energy device

where conventional tidal turbines may struggle. The gentle rotation also minimizes harm to marine life, allowing fish and other organisms to pass through safely, which is a significant ecological benefit. ASTECs tend to have lower noise emissions and relatively straightforward maintenance compared with more complex turbine systems. However, they are generally less efficient at extracting energy per unit area of tidal flow, and their larger physical footprint can limit deployment in constrained locations. Additionally, like all submerged systems, they must contend with biofouling, corrosion, and high hydrodynamic forces, which can increase operational costs over time.

Several innovative Archimedes-screw hydroelectric systems have been deployed around the world. In Wales, the Radyr Hydro Scheme in Cardiff uses two 10-m Archimedes screws delivering a combined capacity of 400 kW. In the US, the Meriden/hanover Pond facility in Connecticut offers roughly 192 kW capacity. So far, ASTECs are start with river flows while have not been commercially operating in ocean tidal currents. However, Norwagian compnay Flumill AS does aim to deploy their "helix turbine" technology in tocean-stream settings [1].

In the future, Archimedes screw tidal energy converters are expected to evolve through modular designs, improved materials, and optimized screw geometries to enhance efficiency and reduce installation costs. Hybrid systems that combine multiple screws or integrate energy storage could further increase their utility in grid applications. While ASTECs are unlikely to compete with large-scale tidal turbines in terms of high-density power generation, their ecological friendliness, adaptability to low-flow environments, and simplicity make them a promising technology for smaller, environmentally sensitive tidal sites. As global interest in tidal energy grows, Archimedes screw systems may become a complementary component of coastal renewable energy portfolios, particularly in regions where minimizing ecological impact is a priority.

### 4.1.4 Hydrofoil

Hydrofoil tidal energy converters are a unique class of devices that harness the lift force generated by the movement of tidal currents over hydrofoil-shaped blades or wings. Unlike turbines that rotate around an axis, these systems use the oscillatory motion of hydrofoils to drive mechanical linkages or hydraulic systems, which then convert the kinetic energy of flowing water into electricity. Typically, one or more hydrofoils are mounted on a central arm or beam that moves up and down in a motion similar to a dolphin's swimming, driven by the flow of water over the wing's surface, as shown in Fig. 4.4. This motion is then translated through a mechanical or hydraulic transmission system to a generator, either located on the device or onshore via power cables.

The key advantage of hydrofoil-based tidal energy converters lies in their efficiency and ability to operate in relatively low flow speeds. Because hydrofoils generate lift rather than relying solely on drag, they can extract energy more effectively from slower-moving water compared to some turbine systems. Their oscillating nature also means they can often be

**Fig. 4.4** Illustration of underwater hydrofoil TECs

positioned closer to the surface, where tidal currents are strongest, and they typically allow for easier maintenance access, particularly when mounted on floating or semi-submerged platforms. Moreover, these systems tend to have a lower speed and a different interaction profile with marine life, potentially reducing the risk of harm to aquatic fauna compared to spinning turbine blades.

However, hydrofoil converters also come with engineering and operational challenges. The mechanical systems that convert oscillating motion into electricity are complex and must be robust enough to withstand continuous dynamic loading and the corrosive marine environment. Ensuring structural stability and fatigue resistance over long-term deployment is a significant design hurdle. Their novel architecture also means fewer commercial-scale deployments have been realized so far, which can slow down investment and standardization efforts.

One notable example is the bioSTREAM device developed by BioPower Systems in Australia. Inspired by the tail movements of swimming fish such as sharks and tuna, bioSTREAM uses flexible hydrofoil appendages that oscillate with the flow of tidal currents. This motion drives power take-off mechanisms to generate electricity. The design emphasizes low environmental impact and is particularly suited for low-velocity or shallow tidal sites. While still in the early demonstration stages, bioSTREAM illustrates the potential of bio-mimetic hydrofoil approaches to capture tidal energy efficiently, offering advantages such as simpler mechanics, quieter operation, and reduced risk to marine life compared with conventional rotary turbines. Other hydrofoil TECs are similarly focused on oscillating foil mechanisms. The Stingray device, developed by the University of Edinburgh, features oscillating hydrofoils that convert tidal current energy into mechanical energy for electricity generation. Designed with a modular structure, Stingray allows for scalability and ease of maintenance while emphasizing minimal environmental impact. Similarly, the Pulse Tidal Pulse Generator harnesses lift and drag forces generated by oscillating hydrofoils to produce electrical power. Pulse Tidal has tested its devices across multiple demonstration projects to validate performance and reliability, highlighting the versatility of oscillating hydrofoil systems for tidal energy applications. Earlier hydrofoil devices also contributed to the field's development. The Aquanator, invented by Australian engineer Michael Perry, used rows of hydrofoils to generate electricity from slow-moving water currents. The system was connected to the national grid in 2006, demonstrating that hydrofoil arrays could operate effectively in low-flow tidal environments. Although the project was decommissioned in 2008, it provided valuable insights into the design, installation, and performance of hydrofoil-based tidal systems.

Looking to the future, hydrofoil tidal energy converters hold promising potential for expanding the reach of marine renewables into areas with lower or more variable flow conditions, where traditional turbines may not be viable. As the technology matures, improvements in control systems, materials, and deployment strategies could reduce costs and enhance reliability. With their compact footprint and potential for modular installation, hydrofoil systems may also find application in off-grid communities, remote islands, and niche markets seeking clean, predictable, and site-adaptable energy solutions. Their unconventional but innovative approach may become an important complement to turbine-based systems, helping diversify the global tidal energy landscape.

### 4.1.5 Kite

Kite-based tidal energy converters are an emerging technology that harnesses the kinetic energy of tidal currents using tethered underwater "kites" or winged platforms. Unlike traditional horizontal-axis turbines fixed to the seabed, these devices are designed to move dynamically through the water, following a controlled figure-eight or other oscillatory trajectory. The kite's movement across the tidal flow amplifies the apparent water speed over the onboard turbine, allowing it to generate electricity even in areas with relatively low tidal velocities. The energy generated by the turbine is transmitted to a seabed-mounted power take-off system, or in some designs, directly along the tether to shore via subsea cables. This mobility allows kites to cover a larger area of flow, increasing the efficiency of energy extraction compared to stationary devices.

One of the most well-known examples of kite-based tidal technology is Minesto's system, presented in Fig. 4.5. This device features a winged, kite-like structure tethered to the seabed, with a small turbine mounted onboard. The system "flies" through the water in a figure-eight pattern, achieving flow speeds over the turbine up to ten times higher than the ambient current, thereby enabling electricity generation in slow-flowing tidal sites. In 2020, Minesto successfully achieved the milestone of delivering electricity to the Faroes Islands' power grid, and this has been expaned in 2022.

The primary advantage of kite-based TECs is their ability to operate in sites with lower tidal velocities, where conventional turbines may not be viable. The dynamic motion of the kite allows for higher energy capture per unit of tidal flow, and the system typically has a smaller seabed footprint, reducing installation challenges and potential environmental impact. KTECs are also relatively quiet and pose less risk to marine life due to slower-moving blades compared with large rotary turbines. However, these systems face several challenges, including the complexity of controlling the kite trajectory in variable flow conditions, the mechanical wear of moving tethers and joints, and the engineering demands of maintaining subsea structures under high hydrodynamic loads. Additionally, deployment and maintenance operations remain costly and require specialized vessels and infrastructure.

Looking to the future, kite-based tidal energy converters are expected to evolve through increased automation, smarter control systems, and modular designs that simplify maintenance and scalability. Advances in materials, particularly lightweight composites and corrosion-resistant coatings, will further improve durability and reduce operational costs.

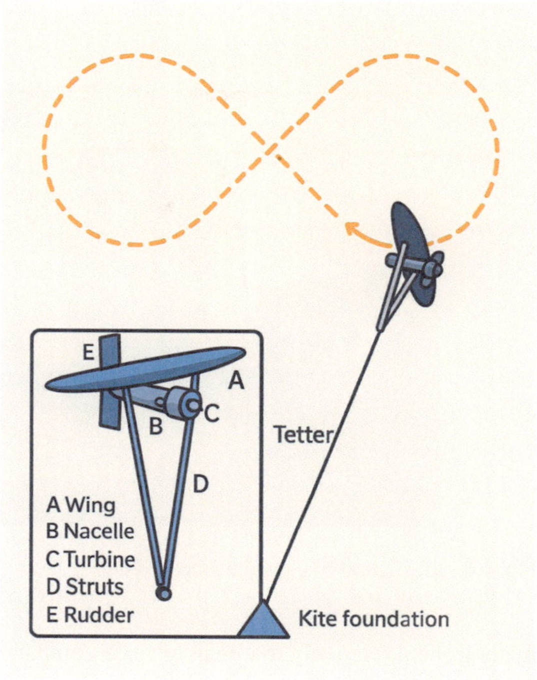


**Fig. 4.5** Illustration of a kite TOC, as per Minesto [2]

As companies like Minesto progress toward multi-megawatt commercial deployments, KTECs could complement traditional tidal turbines, expanding the range of feasible sites for tidal energy generation and helping integrate predictable, low-impact renewable power into coastal energy portfolios.

### 4.1.6 Paddle Wheels

Paddle wheel tidal energy converters are kinetic devices designed to extract energy from tidal currents using large rotating wheels equipped with multiple paddles or blades. The operating principle is straightforward: as tidal currents flow past the paddles, they impart rotational motion to the wheel, which is mechanically coupled to a generator via a gearbox or direct-drive system, converting the kinetic energy of moving water into electricity, as illustrated in Fig. 4.6. The rotation speed of the wheel is generally low, which reduces mechanical stress and minimizes potential harm to marine life. Paddle wheel systems are typically installed in shallow tidal channels, estuaries, or locations with high tidal flow widths, where their large surface area can capture significant energy from broad, relatively uniform currents.

One of the primary advantages of paddle wheel tidal energy converters is their simplicity and robustness. Their slow rotational speed reduces wear on mechanical components,

**Fig. 4.6** Illustration of a paddle wheels TEC

making them relatively low-maintenance compared to high-speed turbines. They can also operate efficiently in low to moderate flow velocities and are less sensitive to sedimentation and debris than other tidal systems. Paddle wheels also have a lower ecological impact due to their slow rotation, allowing fish and other aquatic organisms to pass safely. On the downside, these devices tend to have lower energy density compared with horizontal-axis turbines or kite-based TECs, requiring larger installations to achieve substantial power outputs. Their physical footprint can be significant, potentially restricting deployment in narrow or environmentally sensitive sites. Like all marine systems, they are susceptible to corrosion, biofouling, and hydrodynamic forces that may increase operational costs.

A few paddle-wheel tidal devices include Water Wall Turbine (WWTurbine)'s surface-floating barge at Dent Island, British Columbia—a 28 × 17 m catamaran carrying a 72-ton paddle-wheel turbine, plus Occurrent Power's "Falcon" barge that has a 500 kW powerrate. These efforts reflect the niche exploration of large-paddle systems in tidal/slow-flow environments.

The future of paddle wheel tidal energy converters lies in modularity, improved hydrodynamic designs, and integration with energy storage systems to increase their grid value. Optimizing paddle shapes and materials can enhance efficiency and reduce maintenance needs, while floating or partially submerged platforms could expand deployment flexibility. Although they may not achieve the high power densities of horizontal-axis turbines or kite-based devices, paddle wheel TECs offer a practical, environmentally friendly solution for community-scale or low-impact tidal energy applications. With continued technological development and targeted deployment in suitable tidal channels, paddle wheel systems are likely to remain an important component of the diverse array of tidal energy technologies being explored worldwide.

### 4.1.7 Barrage and Lagoon

Potential energy and water level-based tidal systems, as shown in Fig. 4.7, generate electricity by harnessing the difference in water height between high and low tides. These devices rely on the gravitational potential energy stored in tidal basins or coastal lagoons during tidal cycles, converting it into usable electrical energy. The fundamental principle involves impounding water at high tide behind a structure—such as a barrage or lagoon wall—and then releasing it through turbines as the tide falls (or vice versa), much like conventional low-head hydropower. This class of tidal energy technologies includes tidal barrages, tidal lagoons.

One of the most established and powerful forms of these systems is the tidal barrage, which typically involves a large dam constructed across a tidal estuary. As the tide rises and falls, sluice gates and turbines regulate water flow to capture energy from the change in elevation. These systems can generate a significant amount of power due to the large volumes of water involved and the predictable nature of the tides. Tidal lagoons work on a similar principle but differ in that they are not constructed across natural estuaries. Instead, they are enclosed artificial basins built along the coastline, designed to trap tidal water during high tide and release it during low tide to generate electricity. This approach is seen as more environmentally flexible and scalable, allowing for phased construction and reduced impact on estuarine ecosystems. Though none are yet operational at large

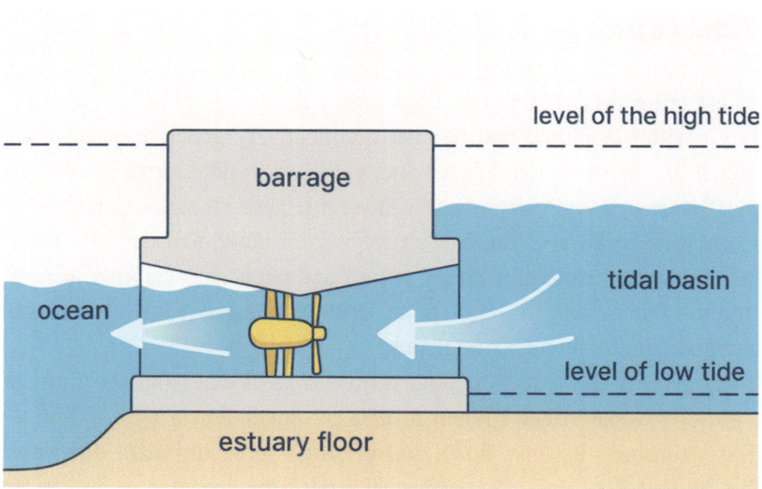


**Fig. 4.7** Illustration of a water level-based tidal system

commercial scale, tidal lagoon projects continue to attract interest as a potentially lower-impact alternative to barrages.

A well-known example is the La Rance Tidal Power Station in Brittany, France, which began operation in 1966 and produces 240 MW of electricity, enough to power around 130,000 homes. A further major example is the Sihwa Lake Tidal Power Station in South Korea, a 254 MW tidal-barrage facility commissioned in 2011 and currently the world's largest operating tidal power plant. A proposed project is the Swansea Bay Tidal Lagoon in the UK, initially planned to deliver around 320 MW capacity, although it has faced delays due to economic and policy hurdles.

Despite their promise, potential energy-based tidal systems face several limitations. The most significant is environmental impact; barrages, in particular, can disrupt estuarine habitats, fish migration routes, and sediment dynamics. Construction costs are extremely high, and project timelines are long, requiring significant upfront investment and political support. Additionally, they are site-specific, requiring large tidal ranges and favorable coastal geography, which limits their global applicability.

Nevertheless, these systems offer compelling benefits. They provide highly predictable, dispatchable energy, meaning output can be planned and controlled based on known tidal cycles. Their capacity factors can be relatively high, and unlike wind or solar, tidal range systems do not depend on weather conditions. With increasing demand for low-carbon baseload generation, especially in coastal nations, interest in tidal lagoons and next-generation barrages is being rekindled. Future potential hinges on developing modular and environmentally sensitive designs, securing regulatory frameworks, and integrating with coastal resilience and flood protection strategies, which could make these large-scale systems not only energy assets but also multifunctional infrastructure [3].

## 4.2 Tidal Farm

Tidal farms are collections of multiple tidal energy converters (TECs) deployed in a coordinated layout within a shared marine site to collectively generate electricity from tidal currents. Much like offshore wind farms and wave farms, tidal farms are designed to harvest energy at scale by taking advantage of favorable hydrodynamic conditions in specific coastal or estuarine locations. Rather than relying on individual devices, tidal farms are essential to achieving meaningful power output and economic viability, especially when considering the high capital and operational costs associated with offshore installations, grid connections, and maintenance logistics.

One of the critical challenges in tidal farm design arises from the farm effect, also known as array or wake effects. When a tidal current interacts with a TEC—such as a horizontal-axis turbine—it slows down and becomes more turbulent downstream of the device. This altered flow, or wake, affects the performance of any downstream TECs, potentially reducing their energy output and increasing unsteady loading and mechanical fatigue. In tightly packed arrays, inefficient wake interactions can significantly compromise the overall farm efficiency, making it crucial to understand and predict these flow interactions during the design phase. Computational modelling and field experiments have shown that the positioning of each device can have a substantial impact not only on its own output but on the collective performance of the entire farm.

Deploying TECs as part of a well-optimized tidal farm enables developers to maximize energy yield, minimize mechanical stress, and reduce environmental impacts. Achieving the best layout involves balancing several factors, including the natural velocity profile of the tidal flow, bathymetry, sediment transport patterns, and ecological sensitivities. Devices must be spaced far enough apart to allow partial recovery of the flow in between, but close enough to ensure that the installation remains compact, cost-effective, and within cable and infrastructure limits. Computational fluid dynamics (CFD) models, tank tests, and in situ monitoring are often used to simulate and refine array configurations. Developers also consider staggered or offset arrangements to reduce wake overlap, and may orient devices to accommodate flow direction changes during tidal reversals.

Beyond technical performance, tidal farm layout also has implications for environmental management, navigation, and spatial planning. Improperly designed farms may disrupt sediment transport or marine habitats, while well-planned farms can be integrated with conservation goals, fisheries, or even coastal protection infrastructure. As regulatory frameworks evolve, developers must also engage with stakeholders to address concerns over marine traffic, visual impact, and cumulative ecosystem effects.

Several tidal energy array designs have been developed and tested to maximize power extraction by carefully managing the interaction between devices and the flow. Because tidal turbines extract energy from moving water, placing multiple devices together—without considering their mutual influence—can reduce overall efficiency due to wake effects. Optimized array layouts aim to minimize wake interference, maximize flow recovery, and increase energy yield per unit area.

One widely studied approach is the staggered or offset array, where turbines are placed in multiple rows with downstream turbines positioned in the gaps between upstream ones. This layout allows the downstream devices to access faster, less turbulent water that has partially recovered from the wakes of upstream units. The spacing between devices is crucial: both the streamwise spacing (distance between turbines in the direction of flow) and the lateral spacing (side-to-side distance) must be carefully chosen. Typically, HATs turbines need to be spaced 4–8 diameters apart downstream (or more [4]), although the optimal configuration depends on the site's flow characteristics and turbine size.

Another strategy involves vertical staggering in locations with significant vertical shear in current speed. In these cases, turbines can be deployed at different depths to take advantage of velocity gradients, such as in a two-tier or multi-layer array, particularly in deeper channels. This design maximizes the use of available kinetic energy throughout the water column and reduces direct wake overlap.

Some innovative designs use channel-constrained arrays in natural or artificial channels where flow is concentrated. By lining turbines along the sides or bed of a channel—such as in the MeyGen project in Scotland's Inner Sound—developers can create a "funneling effect", accelerating the flow between the devices and increasing the power density of the site. In certain high-flow environments, Venturi-ducted arrays have also been proposed, where turbines are housed in structures that accelerate the flow and channel it more efficiently through multiple devices.

## 4.3 Tidal Energy Converters in Seas

The modern era of sea-based tidal stream technology began in 2003 with the installation of the SeaFlow prototype off North Devon, England. This 300 kW seabed-mounted horizontal-axis turbine featured a two-bladed rotor about 11 m in diameter, mounted on a monopile-like support structure. The design allowed the turbine to be raised above the water surface for maintenance and included 180-degree blade pitch control so it could operate in both ebb and flood tides. Although it was not connected to the grid, SeaFlow proved that tidal stream energy could be harnessed at sea with mechanically robust equipment.

In 2008, the SeaGen project in Strangford Narrows, Northern Ireland, marked the first large-scale commercial tidal stream installation. With a rated capacity of 1.2 MW, SeaGen used two 0.6 MW axial-flow rotors mounted on arm-like extensions from a central monopile. Like SeaFlow, the blades could pitch through 180 degrees, allowing them to capture energy from both directions of the tide without reorienting the structure. The system also incorporated a triple-stage gearbox to increase rotor speed for efficient generator operation. SeaGen remained operational until 2019, delivering around 11.6 GWh over its lifetime and demonstrating high reliability, with the ability to raise the rotors above the water for servicing.

The European Marine Energy Centre (EMEC), established in Orkney, Scotland, has since become the world's leading test facility for tidal stream devices. Starting in 2011 at its Fall of Warness site, EMEC hosted a succession of prototypes from different manufacturers. These included Alstom/Tidal Generation Ltd's DeepGen seabed turbine, Atlantis' AR1500 horizontal-axis design, and Scotrenewables' SR2000—a 2 MW floating twin-rotor platform installed in 2016. The SR2000 demonstrated that floating tidal devices could achieve high energy yields, producing more than 3 GWh in its first year while withstanding challenging sea states.

In 2016, Scotland's Pentland Firth became home to the MeyGen project, the largest operational tidal stream array in the world. Developed by SIMEC Atlantis Energy, MeyGen Phase 1A deployed four seabed-mounted horizontal-axis turbines, each rated at 1.5 MW and manufactured by Andritz Hydro Hammerfest and Atlantis Resources. Together they delivered 6 MW to the UK grid, and the site is being expanded in phases toward a planned capacity of nearly 400 MW. MeyGen's seabed-mounted design offers high stability in strong tidal flows, while using proven horizontal-axis rotor technology to maximize efficiency.

Floating tidal technology took a major step forward in 2021 with the launch of the Orbital O2 turbine at EMEC's Fall of Warness site. This 72-m floating platform carries two 1 MW horizontal-axis rotors mounted on retractable legs, each with 20-m blades giving a total swept area of over 600 $m^2$—the largest in the sector. The retractable design allows the rotors to be lifted clear of the water for maintenance without the need for heavy-lift vessels. The O2 is held in place by a four-point gravity-based mooring system and uses 360-degree blade pitch control to capture energy from both tidal directions without rotating the platform. Since entering service, it has provided power to around 2000 homes and supplied electricity for onshore hydrogen production.

In Wales, the DeltaStream device was deployed in 2015 in Ramsey Sound as a demonstration of a gravity-based support structure for tidal turbines. The 400 kW horizontal-axis turbine was mounted on a triangular base that rested on the seabed without the need for piling. Although it was decommissioned in 2016 after technical issues, the project demonstrated the potential for modular gravity-base systems that could simplify installation and removal in future tidal arrays.

## 4.4 Key Challenges

### 4.4.1 Location Limitation

TECs face inherent geographical limitations that restrict where they can be deployed, and these constraints largely explain why the technology is less widespread than offshore wind or wave energy. Unlike wind or wave resources, which are relatively abundant over large offshore areas, tidal stream energy requires sites with fast and predictable water flows. For TECs to be economically viable, peak current speeds generally need to exceed 2–2.5 m per second. Such velocities are typically found where tidal water is concentrated through narrow channels, straits, or around headlands. Globally, these high-flow zones are scarce and often geographically isolated. Notable examples include the Pentland Firth in Scotland, the Bay of Fundy in Canada, and the Raz Blanchard in France. Because these zones are limited in number and size, they represent the principal "hotspots" suitable for tidal stream energy, in contrast to offshore wind or wave energy, which can be deployed across much wider areas of open sea.

Water depth is a critical design and operational consideration for TECs. Devices must be installed at a depth that provides enough clearance for vessels above while remaining shallow enough to allow safe and efficient maintenance. For seabed-mounted turbines, this often means a target range of roughly 20 to 50 m. Shallow waters may limit the ability to capture full tidal energy, while very deep waters complicate installation, increase mooring or foundation costs, and make maintenance operations more challenging. Floating tidal platforms can extend operational depths, but they introduce additional mooring, stability, and cabling complexities, and are generally suited only for sites where seabed conditions are unsuitable for fixed installations. Thus, depth both limits where turbines can be securely installed and affects the choice of device type.

Seabed conditions are equally important: areas with extremely soft sediments, unstable slopes, or high geological variability can compromise the foundations of tidal devices. In some cases, seabed-mounted turbines require rock or consolidated sediments to anchor monopiles or gravity-based structures, whereas floating platforms offer more flexibility but introduce additional mooring and stability considerations. Environmental and operational constraints further reduce the pool of potential sites. TECs must avoid shipping lanes, areas of heavy fishing activity, and regions with sensitive marine habitats. The interaction of turbine structures with marine life, sediment transport, and tidal flow patterns requires careful environmental assessment. Mitigation strategies, such as spacing turbines to reduce hydrodynamic interference or selecting locations outside critical habitats, are essential but can limit the size of deployable arrays. Additionally, logistical access for installation, maintenance, and grid connection often dictates site selection, particularly in remote or harsh offshore environments.

### 4.4.2 Sediment Transport and Seabed Interaction

Tidal energy converters are inherently linked to the seabed, either through fixed foundations or mooring systems, and their operation directly interacts with sediment dynamics. Strong tidal currents transport sediments in varying amounts depending on local bathymetry and flow speed, which can cause both scour and deposition around turbine foundations. Scour (the erosion of seabed material at the base of structures) can undermine stability, leading to tilting or excessive stress on monopiles, gravity-based foundations, or anchors (as shown in Fig. 4.8). In high-flow channels, where currents exceed 2–3 m per second, this effect is particularly pronounced. To mitigate this, protective measures such as rock armoring, geotextile mats, or scour collars are often installed around the turbine base. These armoring solutions stabilize the seabed, reduce erosion, and ensure long-term structural integrity without significantly interfering with local sediment transport.

The interaction with sediment is not limited to the seabed concerns. Abrasive particles carried in the flow can cause significant wear to moving components, including turbine blades, bearings, and gearboxes. Even well-sealed turbines are not immune to microscopic erosion over time, which can reduce efficiency and increase maintenance needs. Compared to offshore wind turbines, which only experience sediment movement at their submerged monopile bases, or wave energy converters, which are often partially floating, tidal devices are fully immersed and subjected to high-velocity particulate flow, creating a unique operational challenge.

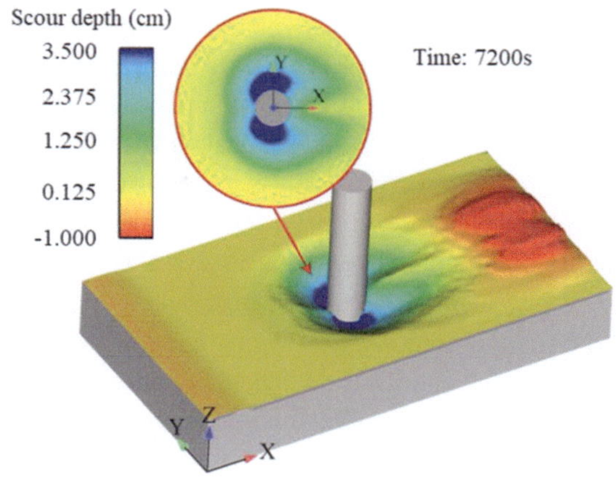


**Fig. 4.8** Scouring around a structure installed on the seabed [5]

Sediment transport also has ecological implications. Changes in flow patterns caused by turbine arrays can alter deposition zones, potentially affecting benthic habitats and nutrient cycles. This means that careful site-specific environmental assessment is essential not only for regulatory compliance but also to ensure that turbine operation does not unintentionally exacerbate erosion or disrupt sensitive seabed ecosystems. In some cases, arrays may need to be spaced or oriented to minimize flow disturbance, which can constrain overall energy yield and site layout flexibility.

In addition, the seabed itself must be suitable for installation. Soft, unconsolidated sediments pose problems for anchoring and foundation stability, while highly uneven or rocky seabeds complicate installation and increase costs. Floating platforms can reduce seabed dependency but introduce additional mooring challenges, especially in channels with high sediment load. The combination of structural, mechanical, and environmental interactions with sediment makes seabed management a critical factor in the long-term success of tidal energy projects.

### 4.4.3 Grid Integration and Variable Power Density

Tidal energy is highly predictable, yet its power output is inherently variable over short timescales due to the cyclical nature of tides. Each tidal cycle produces peak currents for only a few hours, separated by periods of slack water where flow speeds drop near zero. Consequently, the instantaneous power delivered by tidal turbines fluctuates dramatically. Despite being forecastable days in advance, this variability poses challenges for grid integration, particularly for regions seeking to balance supply and demand without relying heavily on storage or backup generation. One solution is to combine tidal energy with complementary renewables, such as offshore wind, wave, or solar. By integrating resources with different temporal patterns, the overall energy supply can be smoothed, reducing the impact of slack tides and maintaining more continuous power delivery.

The characteristic "pulsed" nature of tidal power can also lead to sharp ramping events at the beginning and end of peak tidal flows. Large arrays feeding the grid simultaneously may create surges in electricity injection, requiring careful management through grid controllers, reactive power compensation, or energy storage solutions. Hybrid systems combining tidal and other renewables can help mitigate these sharp fluctuations. For example, offshore wind farms can supply energy during slack tides, while solar PV can contribute during daylight hours. Coordinated operation and smart grid technologies can optimize the combination, balancing energy production with grid demand.

Grid connection infrastructure is another challenge. Tidal projects are often located in remote or offshore channels far from major population centers. This requires high-voltage subsea cabling, transformers, and sometimes onshore substations to transmit electricity efficiently. The costs and technical complexity of installing and maintaining these connections are significant, particularly in areas with strong currents and high sediment transport. Strategic planning and the use of modular or phased deployments can help spread costs and allow incremental grid integration while operational experience grows.

Finally, energy storage systems, including batteries, pumped hydro, or even hydrogen production, can complement tidal arrays to address both temporal variability and sharp ramping events. Such storage solutions, combined with predictive tidal models, enable operators to smooth power delivery and provide more reliable grid services. Integrating tidal energy with hybrid renewable systems and storage therefore represents a practical strategy to overcome the inherent variability of tidal streams while maintaining the economic viability and technical resilience of the project.

## Reference

1. https://flumill.com/
2. Lieber L., Nilsen T., Zambrano C. & Kregting L. Optimising multiple multibeam sonars to assess marine life interactions with an underwater kite. In: Proc. 12th European Wave and Tidal Energy Conference (EWTEC 2017), Cork, Ireland (Sep 2017). Available at: https://tethys.pnnl.gov/publications/optimising-multiple-multibeam-sonars-assess-marine-life-interactions-underwater-kite
3. Khojasteh D., Shamsipour A., Huang L., Tavakoli S., Haghani M., Flocard F., Farzadkhoo M., Iglesias G., Hemer M., Lewis M., Neill S., Bernitsas M.M. & Glamore W. A large-scale review of wave and tidal energy research over the last 20 years. Ocean Eng. 282, 114995 (2023). https://doi.org/10.1016/j.oceaneng.2023.114995
4. Ouro P., Ramírez L. & Harrold M. Analysis of array spacing on tidal stream turbine farm performance using Large-Eddy Simulation. J. Fluid Struct. 91, 102732 (2019). https://doi.org/10.1016/j.jfluidstructs.2019.102732
5. Geng F., Yang W., Nadimi S. & Hu Z. Development and prediction of scour around offshore wind turbine monopile foundations in ebb and flow tides. Ocean Engineering 311, 118905 (2024). https://doi.org/10.1016/j.oceaneng.2024.118905

# Ocean Solar Farms

# 5

Ocean solar farms represent a groundbreaking frontier in renewable energy, leveraging the vast expanse of the world's oceans to deploy floating photovoltaic (PV) panels. Unlike land-based solar installations, these floating platforms capitalize on abundant sea space, offering a sustainable solution to meet global energy demands. With the International Energy Agency (IEA) projecting a need for over 14,000 gigawatts (GW) of solar capacity by 2050 to achieve ambitious climate goals, ocean solar farms provide a compelling alternative to terrestrial systems. Unlike wind, wave, tidal energy devices that are giant structure, what makes ocean solar farms particularly interesting is their modularity and potential to minimize costs.

One of the standout features of ocean solar farms is their ability to save land space. As solar panels require larger span of surface area for more energy harnessing, placing them on usage sea space becomes particularly appealing. For example, to achieve the IEA target of over 14,000 GW solar panel installation by 2050, a surface that is larger than England could be required. However, this would be negligible if deployed on the ocean, only requiring around 0.1% of the open sea. This makes ocean solar an attractive option for lands that are mostly already required for human activities or nature reserves.

In addition, floating solar panels benefit from the ocean's natural cooling effect. Solar PV panels perform optimally at lower temperatures, as excessive heat reduces efficiency. The water beneath floating arrays acts as a heat sink, dissipating thermal energy and maintaining lower operating temperatures. Studies indicate that floating solar systems can achieve efficiency gains of around 10% compared to land-based systems, depending on environmental conditions. This enhanced performance translates to higher energy yields, making ocean solar farms a highly effective solution for large-scale energy generation.



L. Huang, *Ocean Renewable Energy Systems*, Synthesis Lectures on Renewable Energy Technologies, https://doi.org/10.1007/978-3-032-10576-9_5

However, there are also many intriguing aspects of ocean solar farms that remain to be fully explored, for example, ensuring structural survivability under the forces of large waves and severe storms, managing surface fouling such as salt accumulation or bird droppings, and the development of hybrid designs that integrate solar panels with other offshore renewable energy systems. Addressing these issues is critical for the practical deployment and long-term viability of floating solar technologies.

## 5.1 Mechanical and Electrical Systems

### 5.1.1 Modularity

Floating photovoltaic (FPV) systems operate using the same types of solar panels commonly seen on land, such as those mounted on rooftops, with the key distinction that they are deployed on bodies of water. Unlike ground-mounted systems, floating solar panels are installed on buoyant structures, known as floats, which are specifically engineered to remain stable on water surfaces and to withstand environmental stresses such as waves, wind, and varying water levels. The modular nature of these floats is central to the adaptability of FPV systems: each float is designed to carry one or more solar panels, and multiple units can be interconnected to form larger, more complex arrays. This modular design not only simplifies manufacturing, transport, and installation but also makes it easier to tailor the configuration to site-specific constraints, such as irregular shorelines or variable water depths.

FPV systems represent a fully scalable renewable energy solution, where the number of solar panels and floats can be precisely adjusted to meet energy demands across different scales—from small island communities to large utility-scale deployments. A panel-and-float modular approach allows an FPV farm to adapt seamlessly to different locations, sizes, shapes, and operational scenarios. The interconnection between floats can also integrate auxiliary modules—such as walkways for maintenance, storage units for energy buffering, or even hybrid renewable modules like small wind turbines or wave energy converters. Considering diverse power generation and site requirements, an entire FPV platform can be custom-designed by combining standardised float modules into the desired footprint, then anchored in place through an appropriate mooring system, as shown in Fig. 5.1. This modularity provides a clear pathway for incremental capacity expansion, making FPV an attractive option for phased investment and long-term adaptation. In addition, the modularity also offers flexibility to combined with other energy systems, such as wind/wave/tidal, or energy storage systems such as battery and on-site production of fuels.

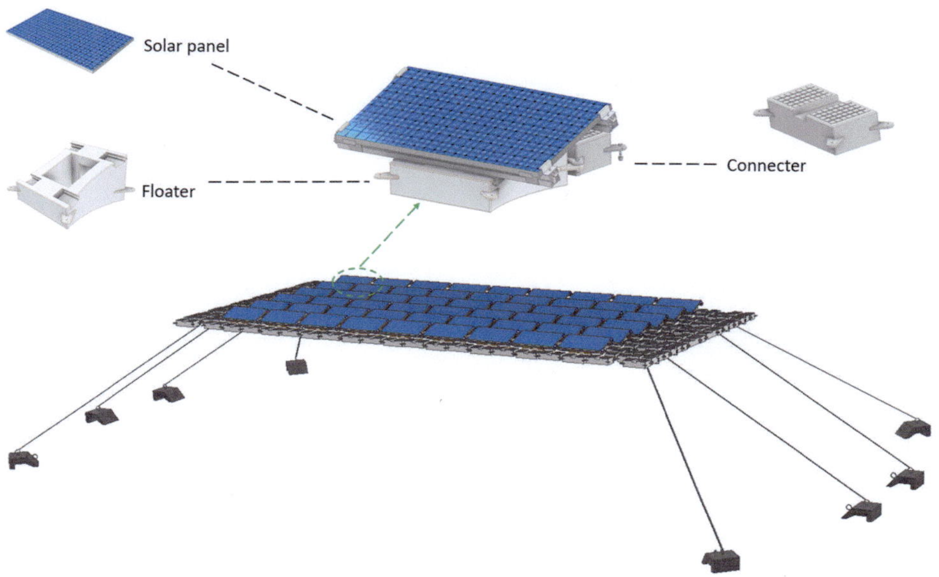


**Fig. 5.1** Illustration of the modularity of a floating solar farm

### 5.1.2 Floater Designs

In order to deploy FPV in open ocean environments, the design, engineering, and operational considerations differ significantly from calm-water projects due to the vastly contrasting environmental conditions. In calm water settings such as lakes, reservoirs, and ponds, the water surface is relatively stable with minimal wave activity and limited wind exposure. These tranquil conditions allow for simpler structural requirements. In contrast, ocean-based FPV systems are exposed to significantly harsher conditions. They must endure constant wave motion, tidal fluctuations, and strong, often unpredictable wind loads. The floatation system must be engineered for durability and flexibility, capable of responding to dynamic loads and maintaining integrity during storms or swells. In this context, various floater systems are being developed to suit ocean applications.

There is a variety of floater designs, representing a major aspect of innovative efforts for marine applications. Figure 5.2 presents common examples. The most common type of float currently used is based on high-density polyethylene (HDPE). They are buoyant, UV-resistant, and chemically stable. For ocean use, they are reinforced with thicker walls, marine-grade additives for salt resistance, and modular interlocking joints that allow flexibility under wave motion. This application has shown effectiveness in open waters with limited wave height and low tidal range, but has not been applied in higher seas.

In higher-energy marine zones, pontoon-style structures are used. These are often made from marine-grade aluminium or steel, sometimes combined with foam or HDPE for

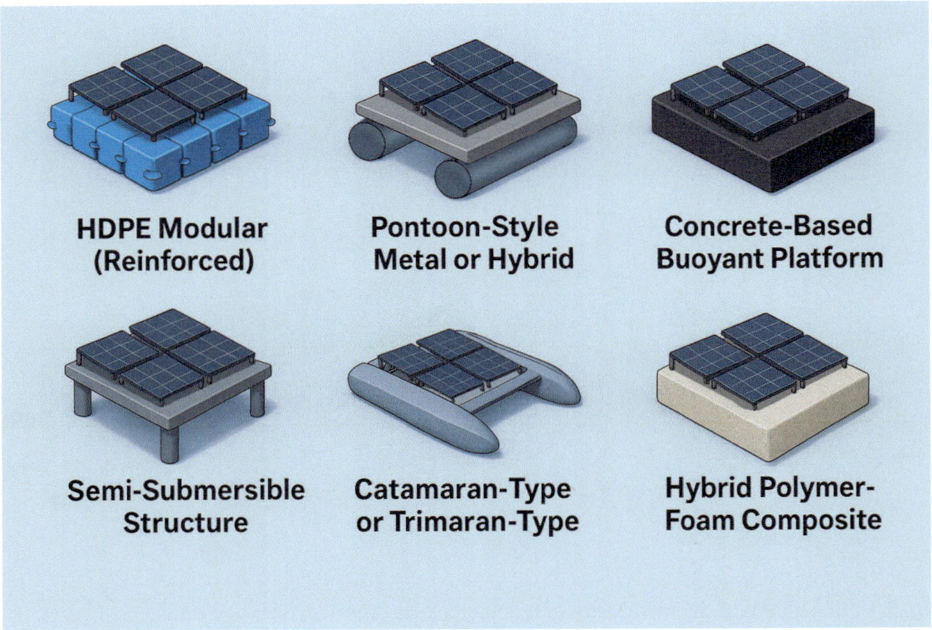


**Fig. 5.2** Illustration of different types of floaters for floating solar farms

buoyancy. They have sufficient structural rigidity to support high ocean loads, and are often applied in combination with corrosion-resistant coatings or sacrificial anodes.

Inspired by offshore oil and wind infrastructure, prestressed or lightweight concrete platforms provide exceptional durability and mass stability. They are heavier and less susceptible to wave-induced oscillation, inherently resistant to corrosion and UV, and deal for large-scale installations in moderate to high wave energy sites. Concrete floats may have hollow chambers or use expanded concrete mixes to ensure buoyancy.

Semi-submersible structures are designed to remain partially submerged, minimizing exposure to wave forces. Similar in concept to floating wind turbine bases, they use multiple floatation columns (columns/pontoons) connected by horizontal beams, coupled with ballasting systems to maintain equilibrium. Semi-submersible designs are ideal for deep-water or high-swell environments, though they are more complex and expensive.

Catamaran-type or trimaran-type floating frames feature two or three hulls (floatation pontoons) joined by a frame, resembling marine vessels. They are a novel type of floats designed to ride waves and minimize damage. However, these are currently still in the research and testing stage and have not proven effective when waves are coming from oblique directions. Foam-based materials could be promising as FPV's floaters, such as the Extruded Polystyrene Foam (XPS), which offers superier strength and buoyance; examples include the Styrofoam produced by DuPont. Some newer systems use foam-core composites or polymer-foam hybrids, combining the lightweight buoyancy of foam, the protective, structural shell of polymer or fiberglass, which resists both corrosion and biofouling.

Film-based floats are an emerging and innovative approach in the field of floating photovoltaic (FPV) systems, particularly in offshore and semi-offshore environments. Unlike conventional rigid or semi-rigid floating platforms, film-based FPVs use thin, flexible membrane structures as both floatation and mounting surfaces for solar panels. The membrane remains in direct contact with the water, allowing for effective thermal management and efficient use of materials. The lightweight, low-profile design reduces both cost and visual impact, while the membrane's flexibility allows it to adapt to wave motion, potentially increasing resilience in dynamic water conditions. However, film-based FPV systems face several technical challenges. Long-term durability is a key concern, as the membrane is continuously exposed to UV radiation, wave action, and potential biofouling from algae or marine organisms. Maintaining the flatness and proper orientation of solar panels on a deformable surface also presents engineering difficulties. The company Ocean Sun is a well-known company for its film-based FPV design.

### 5.1.3 Connections Between Floaters

Modular FPV platforms rely on interlocking mechanisms to link individual floats and form a continuous, stable base for solar panels. The choice of interconnection type depends on the expected environmental conditions, the scale of deployment, and the desired balance between rigidity and flexibility. The most common types of inter-floater connections are illustrated in Fig. 5.3, which includes: (a) Pin-and-Socket Connections: These are mechanical joints where one float has protruding pins or pegs that insert into sockets or recesses on the adjacent float. The connection is often reinforced with locking pins or clips. This method is widely used in calm water FPV systems for its simplicity, ease of assembly, and reconfigurability; (b) Flexible Couplers (Hinges or Straps): In environments where the floating array is subject to wave motion or varying water levels, flexible couplers—such as rubber hinges, synthetic straps, or fabric-reinforced joints—are used. These allow relative movement between floats, reducing stress concentrations and accommodating curvature or slope in the array. They are particularly useful in semi-sheltered marine environments; (c) Bolted or Clamped Connectors: In more rigid configurations, floats may be mechanically fastened using stainless steel bolts, brackets, or clamps. This approach offers high structural integrity but can limit flexibility and is more susceptible to fatigue if not properly engineered; (d) Slide-Lock or Rail Systems: Some commercial systems use slide-in rails or track-based connectors, enabling fast deployment and strong linear alignment. These are efficient in factory-assembled units or large-scale roll-out projects.

The connection between the floating array and the mooring system is essential for maintaining station-keeping under the influence of wind, waves, and current. These connections must allow controlled movement while transmitting loads to the anchors.

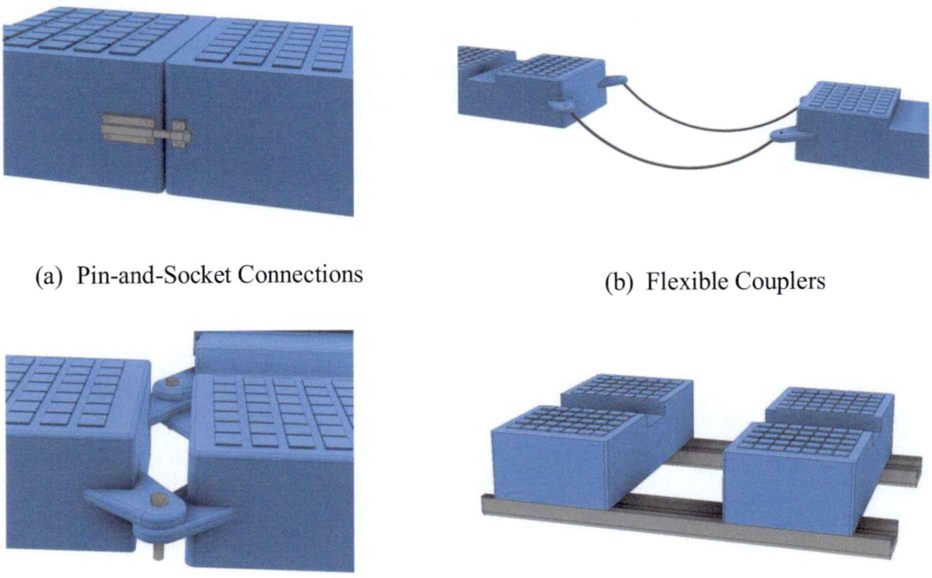

(a) Pin-and-Socket Connections

(b) Flexible Couplers

(c) Bolt

(d) Rail

**Fig. 5.3** Illustration of a few common interconnections between the floaters of FPV

Common float-to-mooring connection types, as shown in Fig. 5.4, include: (a) Chain or Rope Loops with Shackles: A widely used method where marine-grade chains or ropes are attached to dedicated mooring eyes or reinforced lugs on the floats, using metal shackles or D-rings. These provide flexibility and durability and are relatively easy to inspect and replace; (b) Articulated Joints or Universal Couplings: In high-energy marine environments, articulated joints allow multi-directional movement between the float and the mooring line. These reduce bending and shear stress at the connection point and improve survivability during rough conditions; (c) Sliding Anchors (Tension-Leg or Catenary Systems): Some systems use tension-leg moorings or catenary lines where the connection point can slide or pivot as the float rises and falls. This reduces stress concentrations and supports stable vertical motion in tidal or wave-active areas; (d) Hydraulic Dampeners or Elastic Ropes: To reduce dynamic loads and fatigue, elastic or dampened connectors may be used between the float and the mooring line. These absorb wave energy and cushion the system against jerking motions, particularly in exposed waters.

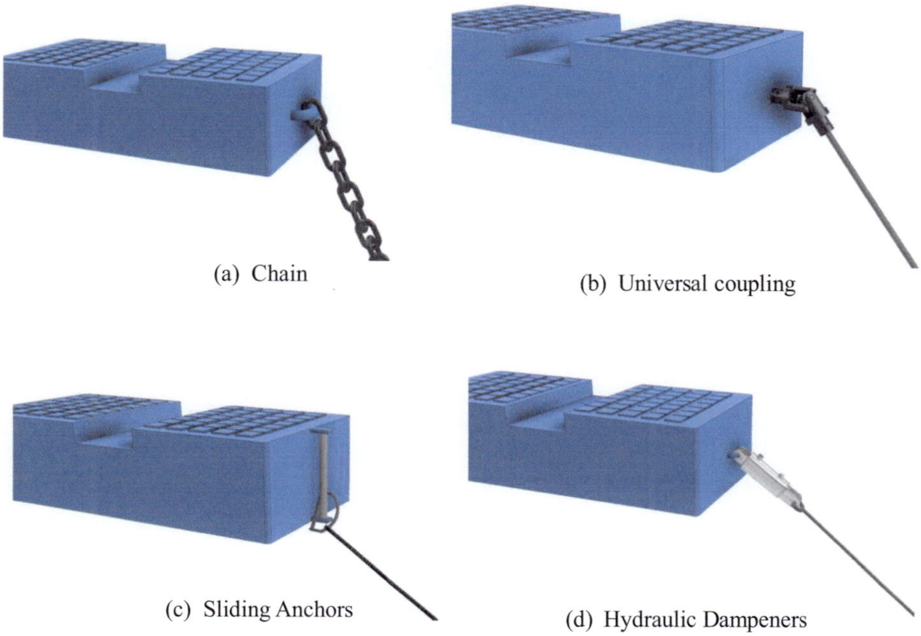

(a) Chain (b) Universal coupling (c) Sliding Anchors (d) Hydraulic Dampeners

**Fig. 5.4** Illustration of common connections between floaters and moorings of FPV

### 5.1.4 Mooring

The mooring system in a floating solar farm is responsible for maintaining the stability and position of the floating platform while allowing it to respond safely to environmental forces such as waves, currents, and tides. In offshore environments, the mooring system is designed not only to keep the platform in place within predetermined offsets but also to accommodate vertical and horizontal motions caused by tidal fluctuations and wave action. A well-engineered mooring arrangement ensures the floating solar array remains operational without imposing excessive mechanical stress on the structure or electrical connections.

Among the various configurations available, the spread mooring system is the most commonly adopted for floating PV platforms. This system involves multiple mooring lines arranged around the perimeter of the platform, each connected to seabed anchors, providing stability from multiple directions and helping to distribute forces evenly under varying wind and wave conditions.

Two primary mooring line profiles are used in offshore floating solar systems: the catenary and taut leg configurations, as shown in Fig. 5.5. In a catenary system, part of the mooring line rests on the seabed in static conditions, with the line's own weight providing flexibility to absorb movement. This design is particularly suited to relatively shallow

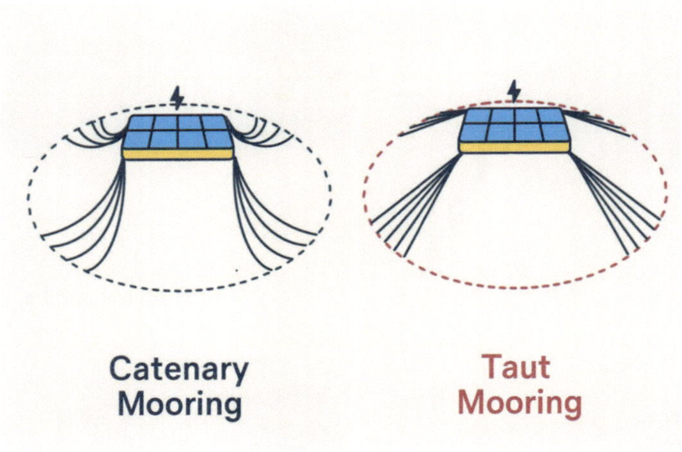


**Fig. 5.5** Illustration of catenary and tout mooring systems

waters (less than 500 m depth) and is cost-effective due to its simplicity and proven reliability. In contrast, taut leg systems maintain constant tension between the floater and seabed anchor, with no slack line lying on the seabed. This arrangement requires less line length but achieves flexibility mainly through the stretching of the mooring material. Taut leg systems are typically deployed in deeper waters, where catenary designs would require impractically long and heavy chains. However, in shallower waters, taut systems may be overly rigid, potentially subjecting the platform to high line tensions during extreme sea states.

The choice of mooring materials and configurations is heavily influenced by water depth and environmental loading conditions. Common setups include all-chain lines, chain–wire–chain combinations, and chain–rope–chain designs, each selected to balance strength, durability, and cost. For example, in depths of less than 500 m, all-chain or chain–wire–chain profiles are often optimal, whereas in deeper waters, chain–wire–chain systems may be more cost-effective due to the reduced weight of wire rope compared to heavy chain. Anchoring systems also vary depending on site conditions and can include deadweight anchors, driven piles, drag embedment anchors (DEA), suction piles, gravity-installed anchors, and vertically loaded anchors (VLA). Each type is chosen based on seabed composition, load requirements, and installation feasibility, ensuring secure and lasting station-keeping.

A robust mooring design process involves calculating expected environmental loads—such as wind, waves, and currents—and applying safety factors to ensure adequate strength and fatigue life over the system's intended service period. This requires assessing not only peak loads but also the long-term cyclic stresses that can cause fatigue in mooring components. In recent years, dedicated mooring design software such as OrcaFlex, MOSES, and AQWA has become integral to simulating performance under a range of sea states, enabling engineers to optimise line lengths, materials, and anchor types before installation. There is also growing interest in innovative mooring technologies, including elastic mooring lines and adjustable tensioning systems, which can enhance survivability and reduce stress during extreme events.

### 5.1.5 Electric Components

The electrical systems in a floating solar farm are integral to converting captured solar energy into usable power and ensuring safe, efficient, and durable operation under challenging marine conditions, as illustrated in Fig. 5.6. At the heart of the system are the photovoltaic (PV) modules mounted on the floating structure, which convert sunlight into direct current (DC) electricity. This DC output is routed via DC secondary cables to combiner boxes that aggregate the power from multiple panel strings, reducing the number of transmission lines needed. From the combiner boxes, DC main power cables carry the electricity to central inverters, where it is converted into alternating current (AC) for transmission to the grid. The system's configuration is typically optimised for large-scale offshore generation, given the economics and complexity of marine deployment, and the layout is tailored to balance efficiency, safety, and ease of maintenance.

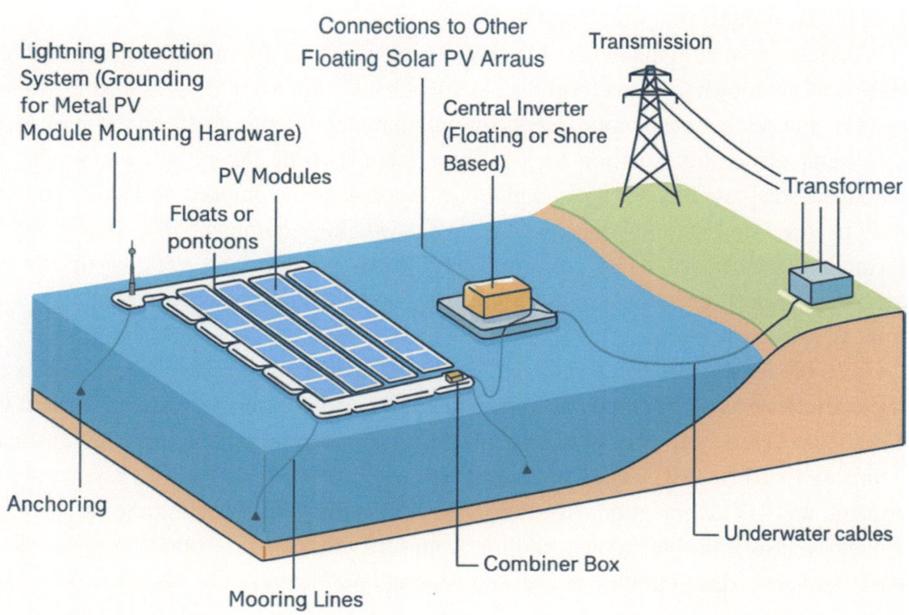


**Fig. 5.6** Illustration of the electrical system of FPV [1]

Cable management is a critical element of electrical design in floating solar farms, as cables must remain secure, protected, and functional despite constant exposure to UV radiation, salt spray, movement from waves, and potential mechanical stresses. On-platform cable trays, clamps, and glands prevent wear and water ingress, while submarine "array" cables—similar to those used in offshore wind farms—connect the floating system to the onshore grid. These submarine cables are often supplied by specialised manufacturers and are installed with protective sheathing and burial or armouring to guard against abrasion, marine growth, and anchor or trawl damage. Onshore, cables are typically routed underground or via overhead lines in compliance with local electrical standards.

The earthing (grounding) system is particularly challenging in marine environments. Offshore, the usual land-based grounding methods are impractical, so underwater grounding electrodes are deployed. Their performance depends on water resistivity, which varies with depth, temperature, and salinity. To ensure effective lightning protection and fault clearance, the grounding system must be carefully designed and periodically inspected for corrosion or degradation. Complementing this is a robust lightning protection strategy, as offshore floating platforms are highly exposed; surge protection devices and marine-grade conductors help prevent damage to sensitive electronics.

For operational safety, a rapid shutdown or emergency shutdown system is often integrated. This system can deactivate sections or the entire array instantly during maintenance, emergencies, or fire hazards, sometimes triggered automatically by abnormal temperature rises. Such safety measures are critical in remote offshore locations, where rapid physical intervention may not be possible.

While essential components such as PV modules, cables, grounding systems, combiner boxes, and shutdown systems form the electrical backbone, many designs also incorporate desirable but non-essential components directly onto the floating platform to improve efficiency and reduce transmission losses. These include string inverters, transformers, and switchgear integrated into dedicated platform sections or in compact "full skid" units that combine inverters, step-up transformers, and switching equipment in a pre-assembled, marine-hardened package. By converting DC to AC offshore and stepping up the voltage before export, these units minimise energy loss over long submarine cable runs and simplify grid connection.

Given the harsh marine environment, corrosion prevention is a constant priority across all electrical systems. PV module frames, fasteners, and mounting points are treated with thick oxide layers, galvanisation, and insulating coatings to prevent galvanic corrosion. Connectors such as MC4 units are selected in marine-grade variants to resist saltwater intrusion and UV degradation. Regular inspection and maintenance regimes are essential to identify insulation breakdown, fouling from bird droppings, or other damage that can cause hot spots, short circuits, or reduced generation efficiency.

In sum, the electrical systems of a floating solar farm are a blend of proven PV technology adapted for the unique demands of the marine environment. The design must ensure not only efficient energy generation but also long-term resilience against salt, humidity, movement, and biological fouling—challenges that make robust engineering and preventative maintenance as important as the choice of components themselves [2].

## 5.2 Ocean-Based FPV Projects Around the World

As of 2025, ocean-based FPV sector is in an early but accelerating phase: transitioning from research and pilot demonstrations to initial commercial deployments.

The technology is still nascent, with most projects in testing or small-scale phases. Harsh ocean conditions have historically limited progress, but recent innovations, like wave-resistant platforms and modular designs, have enabled survival in waves of several meters high. Successful installations demonstrate viability, but full commercialization faces hurdles like high costs, regulatory approvals, and environmental concerns (e.g., impacts on marine life or fishing). Governments in Europe, Asia, and elsewhere are supporting via grants and auctions, with hybrids gaining traction such as offshore solar-wind-hydrogen systems.

In this growing section, many innovative companies are developing various prototypes based on their concepts and launching testing and demonstration projects around the world. SolarDuck, a Dutch-Norwegian startup, is a pioneer in the offshore FPV space with a uniquely engineered solution designed specifically for open-sea conditions. Unlike typical flat-floating platforms, SolarDuck's design elevates solar panels on triangular, rigid frames that sit above the water, mimicking a truss bridge structure. This geometry enables the platforms to remain stable during high waves and heavy winds, significantly reducing saltwater exposure and improving system longevity. The company's early pilot in the Netherlands demonstrated the platform's resilience and scalability, and in recent years, SolarDuck has entered high-profile collaborations, including a partnership with RWE to deploy the Merganser project in the North Sea. They also have a presence in Japan through a project in Tokyo Bay, reinforcing their commitment to expanding in both European and Asian markets. SolarDuck's long-term ambition is to integrate FPV arrays with offshore wind farms to create hybrid energy parks that maximize ocean space efficiency and grid integration.

Ocean Sun, a Norwegian company, has deployed innovative circular membrane-based FPV systems designed specifically for rough offshore conditions. Their demonstration projects in Norway and the Philippines utilize a patented thin membrane floating platform that supports solar panels while minimizing structural weight and improving survivability in waves. These systems have passed rigorous testing at SINTEF Ocean labs and received certification from DNV, paving the way for commercial expansion. Meanwhile, the Netherlands and France are also exploring marine FPV through pilot projects, often in synergy with existing ports or coastal energy infrastructure.

Solar2Wave, based in Indonesia, is developing FPV systems tailored for offshore energy islands and island territories. They provide energy to remote residents who lack access to electric grids. Solar2Wave's innovative concept uniquely integrates floating solar panels with an external breakwater, creating a hybrid structure that shields itself from the ocean's dynamic harshness. Instead of a standalone PV array, their design positions a breakwater—typically a pontoon or concrete barrier—immediately upwave of the solar modules. This strategic placement significantly reduces wave forces and wind loads on the panels, enhancing overall stability and energy capture.

XolarSurf, a new entrant into the ocean solar field, is positioning itself as a designer of next-generation, surfable floating solar platforms. Although still in the pre-commercial phase, the company is experimenting with biomimetic designs that imitate marine species' movement to increase platform resilience. XolarSurf focuses on mobility and modularity, proposing swarms of interconnected FPV rafts that can reposition themselves in response to weather conditions or energy demand. Their early-stage work is closely tied to academic research institutions and maritime engineering labs, particularly in northern Europe. While commercial projects have not yet been launched, XolarSurf's innovative vision could add a new dynamic to the offshore solar space if proven viable.

Fred. Olsen 1848, a Norwegian maritime innovation firm and part of the Fred. Olsen group, has been actively developing Brizo, a wave-tolerant floating solar platform designed for integration into offshore wind farms. Their engineering solution focuses on ensuring structural durability and operational synergy with existing offshore assets. The platform is designed to survive North Sea conditions, including high wave loads and extreme weather events. With strong ties to Norway's offshore oil, gas, and wind sectors, Fred. Olsen 1848 is uniquely positioned to accelerate the commercialization of hybrid offshore renewable solutions. The company envisions a future where solar, wind, and hydrogen production platforms co-exist in multi-use marine zones, enabling year-round renewable output and smarter use of marine space.

In Singapore, the company G8 Subsea is leading the development of a 5 MW nearshore FPV installation in the Straits of Johor, one of the first of its kind to be moored in marine waters with significant tidal flows. This project is using flexible mooring systems and corrosion-resistant components to withstand the harsh marine environment. It is seen as a key testbed for scalability in Southeast Asia, where limited land area makes offshore renewables particularly appealing. Similarly, South Korea has expressed ambition to build large-scale marine FPV farms, leveraging its extensive coastline and engineering expertise in floating structures developed from its offshore oil and gas sector.

## 5.3 Key Challenges

### 5.3.1 Structural Integrity

One of the primary structural integrity challenges for ocean-based FPV systems is the risk of cracks in floaters and failure of connectors, driven by the relentless environmental loadings from wind, waves, tides, and currents. These dynamic forces impose cyclic stresses on the floating platforms, leading to material fatigue over time, where repetitive motions [3] can cause microcracks in the buoyant structures or outright fractures in high-stress areas like module edges. For instance, wave overtopping can exacerbate this by slamming water onto the panels, potentially creating hidden cracks in solar cells and reducing their efficiency, while multi-directional waves amplify shear forces that strain inter-floater connections. Connector failures are particularly acute, as these joints—linking individual floatation modules or attaching to mooring systems—endure constant flexing, which may result in loosening, deformation, or complete breakage if not designed for marine-grade durability. Salt-induced corrosion further compounds these issues, as prolonged exposure to saline environments accelerates degradation of metallic components, such as anchors, cables, and frames, leading to pitting, weakening, and eventual structural compromise. This corrosion is intensified in offshore settings, where UV radiation and biofouling (e.g., marine organism buildup) add layers of stress, promoting fatigue and reducing the lifespan of the system.

Beyond these core challenges, other structural risks include mechanical fatigue at connection points and overall platform instability during extreme weather events. In ocean environments, the perpetual movement from waves and wind creates high-cycle fatigue in joints and moorings, where even moderate loads repeated over years can lead to failure without proper dampening or reinforcement. Biofouling poses an additional threat by increasing weight and drag on the structure, potentially altering hydrodynamics and amplifying loads during storms, while climate change projections suggest heightened fatigue damage from more intense wave patterns in the future. These risks highlight the dilemma of balancing survivability with cost-effectiveness: robust designs for harsh conditions often inflate expenses, making large-scale deployment challenging.

To mitigate these challenges, innovative solutions like fixed and floating breakwaters have emerged as effective protective measures. Breakwaters, positioned upstream or around the perimeter of FPV arrays as shown in Fig. 5.7, dissipate wave energy and act as wind barriers, reducing induced motions and structural loads, allowing the use of lighter, cost-effective floaters adapted from inland designs [4]. Floating breakwaters, in particular, offer flexibility in open seas by absorbing energy without linear cost scaling, making them ideal for expansive farms. Other mitigations include employing flexible connectors with elastic ropes or advanced joint geometries to absorb shocks and minimize fatigue, alongside marine-grade materials (e.g., corrosion-resistant alloys) and regular fatigue life assessments to predict and prevent failures. Comprehensive mooring systems with optimized anchors further limit excessive movement [5, 6] , while anti-biofouling coatings and UV-resistant polymers address secondary degradation risks, paving the way for resilient, economical offshore solar solutions.

The long-term reliability of submerged electrical infrastructure must also be considered. FPV cables submerged in saltwater environments can suffer accelerated insulation degradation, with the potential for copper leaching and environmental contamination. The high corrosion rates characteristic of marine conditions add another layer of complexity to system design, requiring that mooring and cabling strategies be integrated from the outset. Addressing these mechanical and electrical durability challenges in parallel will

(a) Coastal case

(b) Offshore case

**Fig. 5.7** Illustration of breakwaters attached at the wave-coming direction of a floating solar farm

be essential for enabling the safe, scalable, and cost-effective deployment of ocean-based FPV power plants.

### 5.3.2 Mooring Methodology

Designing an effective mooring system for wide-span, ocean-deployed FPV farms introduces complexities far beyond those of moderate-size, freshwater arrays. For MW and GW level of solar farms, the number of mooring lines required is enormous, which is much higher than floating wind, wave, or tidal devices that normally require a few or dozens of mooring lines. There is a lack of methodology to design such mooring systems with numerous mooring lines in the industry. Conventional mooring simulation tools such as OrcaFlex, MOSES, and AQWA typically accommodate only around one hundred moving bodies. Scaling simulations to the thousands of modules present in even a 1 MW FPV array induces high computational demands, making it increasingly difficult to model behaviours of the system, where thousands of floating bodies are coupled and interact with moorings as well as the surrounding wind/water. In this context, there is a pressing need for new mooring design methodologies that can consider the large FPV farms without oversimplifying critical interactions.

In open-water environments, large FPV arrays are exposed to spatially uneven wind, wave, and current loading. Studies have shown that wave-induced forces can increase with the number of array columns, and that the combined effects of wind and waves vary significantly across different positions within the array. Large surface coverage of a utinity-scale FPV on the ocean means that conventional edge-only mooring systems are no longer adequate. Without additional internal mooring points, the interior sections of the solar farm are vulnerable to localised stress and deformation [7], which can compromise both structural integrity and energy production efficiency. Strategically placed internal mooring lines can help to distribute these forces more evenly, but their optimal spacing, orientation, and anchoring method remain active areas of research.

### 5.3.3 Soiling and Cleaning

On ocean-based FPV installations, salt residues are a persistent concern. Salt-laden sea spray and mist settle onto panel surfaces, eventually crystallizing and diffusing sunlight. This significantly lowers energy output and, over time, may also cause corrosion or degradation of panel coatings. Bird droppings are often a major fouling issue on FPV, although in many parts of South Asia the avian activity around sea-based installations is relatively low. That said, salt remains the major contaminant that compromises panel performance [8].

Cleaning FPV systems differs markedly from land-based maintenance. Manual cleaning is both hazardous and impractical on floating arrays, where unstable surfaces and access constraints pose logistical and safety challenges. Moreover, transporting personnel or equipment across water can be costly and weather-dependent. The need for reliable, low-impact cleaning methods is urgent to prevent long-term energy losses and reduce operational expenditure.

One of the most notable innovations addressing this issue small on-panel robotic systems for cleaning [9]. An approach is the water-only robotic cleaner developed by TG hylift, known as the hyCleaner black Solar facelift [10]. Tested on a 500 kW FPV installation in Spain, this battery-powered machine operates via remote control and uses a water-fed rotating brush system to clean panels. Its design minimizes disturbance to the floats and requires no chemical detergents. Edge detection features help prevent accidents by automatically stopping the robot at the boundaries of the floating array [11].

Beyond mechanical systems, several cutting-edge technologies are emerging. Electrostatic cleaning systems, such as those developed by CleanFizz [12], use travelling electrostatic waves to levitate and displace dust and salt particles from panel surfaces. Integrated electrodes beneath the glass work with AI and sensors to trigger fully automated cleaning—no water or moving parts required. This approach could dramatically reduce maintenance needs, particularly in dusty or offshore environments. Other FPV designs are incorporating self-cleaning modules directly into the floats or mounting frames, such as rotating brushes or water-jet mechanisms that operate automatically and reduce reliance on external servicing.

### 5.3.4 Tilting with Waves

In FPV systems deployed in ocean environments, a significant challenge arises from the power loss induced by the pitching motion of solar panels in response to waves. This phenomenon occurs as ocean waves impart constant momentum to the floating structures, causing the panels to undergo periodic oscillations that continuously alter their tilt angle relative to the incoming sunlight [13, 14]. Unlike stationary land-based or calm-water solar installations, where panels can maintain an optimal fixed orientation for maximum irradiance capture, these wave-induced motions disrupt the alignment, leading to sub-optimal energy harnessing efficiency. This dynamic tilting is particularly pronounced in coastal and offshore regions, where waves are a persistent environmental factor, transforming what could be a stable energy generation process into one plagued by variability and reduced output [15].

The mechanisms underlying this power loss are rooted in the rotational movements triggered by waves, which directly correlate with diminished electrical performance. As panels pitch with the waves, the effective angle of incidence for solar radiation fluctuates, scattering or reducing the amount of light absorbed by the photovoltaic cells. This results in a measurable drop in power generation, with the study demonstrating that FPV units in wavy conditions produce significantly lower output compared to their counterparts in still water. An empirical equation derived from published research [16] links the extent of power loss to the amplitude of these rotational motions, providing a predictive tool for quantifying the impact. Such losses not only lower overall system efficiency but also complicate energy yield forecasting, posing economic hurdles for large-scale ocean solar deployments where consistent performance is crucial for viability.

To mitigate these challenges, wave attenuation technologies, such as breakwaters, can be promising, which can dampen the incoming wave energy and minimize the induced motions on FPV platforms. By strategically placing fixed or floating breakwaters around the solar arrays, the amplitude of pitching can be reduced, allowing panels to maintain a more stable tilt angle and thereby preserving higher levels of power output. This approach not only addresses the root cause of motion-related losses but also enhances the structural resilience of the systems against harsh marine conditions. Future designs could further incorporate advanced mooring systems or adaptive tracking mechanisms that compensate for minor oscillations, potentially integrating real-time sensors to optimize orientation dynamically. Overall, these mitigations represent a pathway toward making ocean-based FPV more reliable and economically competitive, bridging the gap between innovative offshore potential and practical implementation.

## References


1. Lee N., Grunwald U., Rosenlieb E., Mirletz H., Aznar A., Spencer R. & Cox S. Hybrid Floating Solar Photovoltaics-Hydropower Systems: Benefits and Global Assessment of Technical Potential. Renewable Energy 162, 1415-1427 (2020).
2. Huang L., Elzaabalawy H., Sarhaan M., Sherif A., Ding H., Ou B., Yang D. Developing reliable floating solar systems on seas: A review. Ocean Engineering 322, 120525 (2025)
3. Wei Y., Zou D., Zhang D., Zhang C., Ou B., Riyadi S., Utama I. K. A. P., Hetharia W., Wood T. & Huang L. Motion characteristics of a modularised floating solar farm in waves. Physics of Fluids 36, 033320 (2024).
4. Yang Y., Mi C., Ou B., Wong A., Duffy J. G., Wood T., Utama I. K. A. P., Chen W. & Huang L. A comparative experimental study on the hydrodynamic performance of two floating solar structures with a breakwater in waves. Solar Energy 284, 113029 (2024).
5. Yang Y., Ren K., Zhou B., Sun S.Y. & Huang L. Wave interaction with multiple adjacent floating solar panels with arbitrary constraints. Physics of Fluids 36, 037121 (2024).
6. Yang Y., Huang L. & Meylan M. H. Wave interaction with multiple floating elastic plates with arbitrary constraints near a sloping beach. Phys. Fluids 37, 032103 (2025).
7. Yang Y. & Huang L. Waves and structural strain induced by a uniform current flow underneath a semi-infinite floating solar coverage. Phys. Rev. Fluids 9, 094804 (2024).
8. Singh N. K. & Sadhu P. K. The investigation of energy and economy for floating solar PV system on saline water. Microsystem Technologies 30, 1079–1086 (2024). oating solar power plants. Renewable and Sustainable Energy Reviews 209, 115094 (2025).

9. Wei Y., Khojasteh D., Windt C. & Huang L. An interdisciplinary literature review of floating solar power plants. Renewable and Sustainable Energy Reviews 209, 115094 (2025).
10. Sánchez Molina, P. Cleaning robot for floating PV. PV Magazine International (8 March 2021).
11. Solar panel cleaning robot hyCLEANER® black SOLAR facelift convinces while cleaning high-efficient floating PV systems on unused water environments. FineEngineering Magazine, 18 March 2021.
12. https://www.cleanfizz.com/
13. Mi C., Verdin P.G., Nair A., Lyu X., Sherif A., Silva D., & Huang L. An experimental study on the hydrodynamic performance of modular rope-mesh floating solar platforms. Ocean Engineering 339, 122175 (2025)-and-cleaning-strategies
14. Ou B., Cerik B., & Huang L. Seakeeping analysis of catamaran and barge floats for floating solar platforms. Ocean Engineering 332, 120970 (2025).
15. Zheng Z., Hu J., Huang Q., Jin P., Yang Y., Huang L. & Zhou Z. Optimization of dual-module floating photovoltaic arrays: Layout configuration and damping mechanisms for enhanced stability and energy performance. Energy 330, 136879 (2025)
16. Huang L., Yang Y., Khojasteh D., Ou B. & Luo Z. Floating solar power loss due to motions induced by ocean waves: An experimental study. Ocean Engineering 312, 118988 (2024)

# 6 Future Perspectives: Hybrid, Integration, and Digitalization

## 6.1 Hybrid Ocean Renewable Systems

### 6.1.1 Floating Solar Intersecting Offshore Wind Farms

The future of offshore renewable energy is increasingly moving toward hybrid systems that co-locate and integrate multiple energy harvesting technologies within a single marine platform. There are hybrid ocean energy systems that use multiple of the above-mentioned energy systems. As shown in Fig. 6.1, with offshore wind energy development accelerates globally, a new opportunity has emerged to enhance energy density and infrastructure efficiency: the deployment of Flloating PhotoVoltaic (FPV) systems in the unused sea space between large wind turbines within an offshore wind farm. This hybrid configuration allows developers to make better use of the vast, underutilized areas between turbine foundations, leveraging the same ocean footprint for multiple forms of clean energy generation.

Offshore wind farms typically have large inter-turbine distances to avoid aerodynamic wake effects and allow for maintenance access. For example, a wind farm with 10–15 MW turbines may have a spacing of around 1 km between adjacent towers, resulting in significant open sea surface between rows and columns. This unoccupied space, which can make up more than 90% of the total area of a wind farm, presents a valuable opportunity for integrating floating solar systems.

Deploying modular floating PV platforms in this space offers multiple benefits. First, it significantly increases the energy yield per square kilometer of ocean area. Secondly, shared infrastructure (including export cables, substations, maintenance vessels, and port facilities) can lower the capital and operating costs of the solar component, enhancing overall project economics. Grid connection costs, which can constitute a major portion of offshore project expenditure, are particularly attractive to share.



L. Huang, *Ocean Renewable Energy Systems*, Synthesis Lectures on Renewable Energy Technologies, https://doi.org/10.1007/978-3-032-10576-9_6

**Fig. 6.1** Illustration of floating solar farms deployed in the empty space between offshore wind turbines

Another key benefit is resource complementarity. Solar and wind resources often have different generation profiles. Solar peaks during the day under clear skies, while wind is frequently stronger at night or during stormy conditions. Combining both sources smooths the total power output, reduces curtailment risk, and improves the utilization rate of the export cable and grid connection. In some locations, seasonal variation also helps: summer solar can offset lower wind production and winter wind can offset lower solar production.

However, co-deployment presents technical challenges. Floating solar platforms must be engineered to withstand offshore wave conditions, salinity, and biofouling—issues that are more severe than in inland or sheltered floating PV applications. Mooring and anchoring strategies must be designed to avoid interfering with wind turbine foundations, cable routes, and vessel navigation. Moreover, energy management systems will be important to coordinate generation forecasts and grid dispatch between solar and wind components.

Another integration strategy involves mounting or tethering WECs directly to the floating foundation of an offshore wind turbine. Several designs have explored this concept, particularly for semi-submersible or barge-type wind platforms that have sufficient surface area and stability. Wave energy devices, such as oscillating flaps, point absorbers, or air-filled chambers (oscillating water columns), can be integrated into the hull or sponsons of the floating foundation, using the platform's inherent wave exposure and structural mass. For example, projects like CORDIS have proposed designs where wave energy components are installed around the periphery of wind platforms, transforming motion into electricity through hydraulic or pneumatic systems.

### 6.1.2 Combined Wind, Wave, Tidal and Solar System

Another concept, as given in Figure, shows how offshore space can be efficiently utilized by combining FPV, offshore wind turbines, and wave/tidal energy systems into one cohesive infrastructure. The floating platform in Fig. 6.2 supports a large floating solar array. Mounted above the solar platform are vertical-axis wind turbines, which are more suitable than horizontal-axis wind turbines in this case due to smaller scales. The vertical-axis design allows for compact placement and omnidirectional wind capture, reducing aerodynamic interference between adjacent units. Surrounding the main platform are long, floating structures resembling wave energy converters or tidal energy converters, such as OWC. These devices extract kinetic and potential energy from the seawater movement. They provide a valuable complement to the intermittent nature of wind and solar resources. Tidal energy converters could also be integrated beneath the platform or nearby, either as seabed-mounted turbines or tethered floating systems. The wave and tidal devices not only provide additional energy, they also protect the floating platform from sea loads like breakwaters, as they turn the water momentum into useful energy, calming the internal water surface around the platform. The NoviOcean's prototypes provide a good example in this domain.

The hybrid configuration offers several technical and operational advantages. Each renewable energy source has a different temporal generation pattern—solar peaks during the day, wind is often strongest at night or in stormy conditions, wave energy can persist

**Fig. 6.2** Illustration of a hybrid offshore energy platform, with additionally the potential to provide clean energy for ships

through cloudy periods, and tidal flows follow highly regular cycles. Together, they form a more balanced and resilient power generation portfolio, smoothing out fluctuations and reducing the need for storage or backup.

## 6.2 Sustainable Marine and Offshore Networks

### 6.2.1 Maritime Decarbonization

As the global maritime sector faces increasing pressure to decarbonise under International Maritime Organization (IMO) regulations and net-zero targets, offshore renewable energy systems are emerging as a novel solution to help ships reduce their reliance on fossil fuels. There are two mainstreams for ships to reduce their carbon prints in operations, which are to change the powering methods to electric or sustainable fuels. In addition, the electricity and fuels used to fuel ships should be produced by clean energy. By strategically integrating electric charging, hydrogen/ammonia refueling, and automated servicing infrastructure into offshore wind, solar, and hybrid platforms, these renewable energy hubs can support the clean transformation of coastal and ocean-going vessels. Fuelling stations in the middle of the ocean will provide more opportunities for ships to overcome challenges to install very large batteries or fuel tanks that are required for coast-to-coast transits.

Also in Fig. 6.2, charging stations, fuel production units, and passing ships are included. One approach involves equipping offshore renewable energy platforms with high-capacity battery charging stations for electric or hybrid-electric ships. Coastal ferries, research vessels, service operation vessels (SOVs), and port tugs operating on short routes can benefit from mid-route charging points, reducing the need for large onboard batteries and enabling more frequent, lighter, and efficient charging cycles. These systems may use wireless inductive charging pads installed at berthing arms or floating docks, allowing contactless power transfer without human intervention, as shown in Fig. 6.3. The deployment of automated charging arms or drones further enhances turnaround speed and reduces labor costs, making clean refuelling as fast and convenient as traditional bunkering.

For longer-range ships such as cargo vessels, offshore renewable platforms can serve as floating green fuel depots, producing and storing hydrogen, ammonia, methanol, or synthetic LNG using on-site wind, solar, wave, or tidal energy. These fuels can be transferred to vessels via automated pumping systems or modular fuel tanks, which can be swapped at sea using robotic arms or submersible exchange systems. In this setup, “energy tender ships” could also shuttle green fuels from offshore production platforms to passing ships, reducing port congestion and extending the operational radius of low-emission vessels, or even transport the fuels back on onshore for industrial decarbonization. Figure 6.4 presents a concept where an offshore renewable energy platform can produce green fuels

**Fig. 6.3** Illustration of a wireless charging system for electric vessels

and store them under the platform, while a robotic system can help the ship to pick up or swap the fuel tanks.

The creation of renewable-powered offshore waystations—effectively "filling stations at sea"—can enable ships to plan their routes more flexibly, reducing the need to carry large volumes of fuel or batteries. This is especially valuable for zero-emission vessels, whose energy storage is still more limited than conventional fossil-fueled vessels. Offshore charging or fueling platforms could be co-located within existing offshore wind farms or marine hybrid energy hubs, creating synergies between maritime logistics and clean energy infrastructure.

### 6.2.2 Fuels as Energy Storage

Offshore renewable energy systems are suitable for producing green hydrogen, ammonia, and synthetic fuels. These fuels, derived through electrochemical and thermochemical processes powered by renewable electricity, are crucial for decarbonizing sectors where direct electrification is not feasible, such as shipping, aviation, heavy industry, and long-duration energy storage.

**Fig. 6.4** Underwater storage of fuels produced by offshore renewable energies, which can be picked up or swapped by ships

One pathway begins with the production of hydrogen via water electrolysis. In this process, electricity from offshore wind or other renewables is used to split water ($H_2O$) into hydrogen ($H_2$) and oxygen ($O_2$) using proton exchange membrane (PEM) or alkaline electrolyzers. Offshore electrolyzers can be installed on floating platforms, at substations, or onshore if electricity is transmitted via subsea cables. In cases of this book, hydrogen may be produced directly offshore to avoid the cost of marine cables for long-distance electricity tranmission and the losses in such high-voltage transmission.

A promising advancement in offshore green fuel production is the development of seawater electrolysis systems, which eliminate the need to supply purified freshwater to offshore platforms. Traditional electrolyzers require deionized or distilled water, but transporting or desalinating water offshore adds complexity and energy consumption. Seawater electrolysis directly uses the abundant saltwater resource beneath floating or fixed platforms, offering a more self-contained and scalable solution for producing green hydrogen at sea.

However, seawater electrolysis presents unique technical challenges. The high salt content, particularly chloride ions, can lead to the production of chlorine gas, catalyst degradation, and electrolyzer corrosion. To overcome this, researchers are developing advanced electrode materials (such as nickel–iron or cobalt-based catalysts) and selective membranes that suppress chlorine evolution in favor of hydrogen and oxygen production. Some systems use pre-filtration and partial desalination as an intermediate step, while others are exploring fully direct seawater electrolysis under variable current profiles. If successfully scaled, these technologies could significantly improve the autonomy and viability of offshore hydrogen production, enabling entirely ocean-based green fuel supply chains.

Once hydrogen is available, it can be used as a fuel itself, or serve as a feedstock for more complex synthetic fuels. For example, green ammonia ($NH_3$) is synthesized by combining green hydrogen with nitrogen (extracted from air) via the Haber–Bosch process. This is especially attractive for the shipping industry, as ammonia is easier to store and transport than hydrogen, has higher energy density, and can serve as a marine fuel with zero carbon emissions at the point of use. Moreover, ammonia can also act as a hydrogen carrier for later reconversion back to $H_2$ near end-use sites.

Another pathway involves producing synthetic hydrocarbons, such as methanol, methane, or even aviation fuels. These are created by reacting green hydrogen with captured carbon dioxide ($CO_2$) in a process called Power-to-Liquid (PtL) or Power-to-Gas (PtG). These synthetic fuels are compatible with existing infrastructure (pipelines, engines, refineries), making them an attractive option for hard-to-abate sectors.

The benefits of offshore-based green fuel production are considerable. Offshore energy resources, such as wind, are more abundant and consistent than many onshore counterparts, offering higher capacity factors and more stable electricity flows for continuous electrolysis. Additionally, using the power offshore avoids grid bottlenecks, allowing excess generation to be converted directly into storable fuels. Integrating fuel production modules directly into offshore platforms or energy islands can also enhance the economic viability of projects in remote marine zones that are otherwise far from transmission infrastructure.

Logistics is another hurdle. Hydrogen is difficult and expensive to compress, liquefy, or convert for transport, requiring specialized pipelines, storage tanks, or chemical carriers like ammonia or liquid organic hydrogen carriers (LOHCs). New infrastructure will be needed to store and move fuels. Additionally, regulatory frameworks and permitting for offshore fuel production are still evolving and often lag behind the pace of technological innovation.

### 6.2.3 Clean Oil and Gas Production

Offshore oil and gas platforms are among the most energy-intensive industrial operations, requiring continuous power for drilling, pumping, compression, monitoring, and accommodation systems. Traditionally, these platforms rely on on-site diesel or gas-fired generators, often powered by fuel extracted from the reservoir itself. While self-contained, this method is both costly and environmentally damaging. Around 5% of all the oil and gas produced is burned on-site just to keep these platforms operating, producing nearly 200 million tonnes of $CO_2$ every year. That is half a percent of the planet's total emissions. As the global energy sector pivots toward decarbonization, there is growing interest in hybridizing offshore hydrocarbon platforms with renewable energy systems to cut emissions and improve operational efficiency.

One of promising strategies involves electrifying platforms using power from nearby offshore wind farms. This can be done via dedicated subsea cables, connecting fixed-bottom wind turbines to oil and gas infrastructure. For example, in the North Sea, the Hywind Tampen project led by Equinor, has deployed 11 floating wind turbines with a combined capacity of 94.6 MW, supplying renewable electricity to the Gullfaks and Snorre oil fields in Norway. This is a significant project towards greening offshore fossil operations. Some feasibility studies are underway in the UK, the Netherlands, and Australia, assessing wind-to-platform electrification either via grid-tied arrays or standalone floating wind systems. The economic incentives are also strong. Operating diesel generators offshore is expensive, with fuel logistics contributing to high operational expenditures (OPEX). Burning fuel that could otherwise be sold represents a lost revenue opportunity. By offsetting fuel use with renewable inputs, oil and gas companies can improve asset economics, extend field life, and reduce carbon taxes or offset obligations, particularly in jurisdictions implementing strict emissions regulations.

Wave energy is also of great potential in this case. Figure 6.5 demonstrates an idea to integrating WECs onto existing structures of existing structures of offshore platform. It features a hybrid mechanical–piezoelectric WEC system. This design features flexible carpets mitigating stress and capturing energy across a wide spectrum of wave frequencies, while the mechanical pistons enhancing power generation. Together, they provide efficiency without sacrificing durability, creating a system that can both withstand storms and steadily supply clean energy.

### 6.2.4 Incorporation with Aquaculture

Aquaculture, the controlled cultivation of aquatic organisms such as fish, shellfish, crustaceans, and algae in marine or freshwater environments, plays a pivotal role in global food security, producing over half of the seafood consumed worldwide and supporting livelihoods for millions. Unlike traditional fishing, which depletes wild stocks, aquaculture involves breeding, rearing, and harvesting species like salmon, mussels, seaweed, and shrimp in enclosed systems, often using nets, cages, or ponds. However, as demand for ocean space intensifies amid climate change and population growth, a compelling opportunity arises: combining aquaculture with offshore renewable energy systems—including wind, wave, tidal, and solar—to create multi-functional marine hubs that generate clean power while farming the sea, fostering economic efficiency, reduced environmental footprints, and resilient coastal communities.

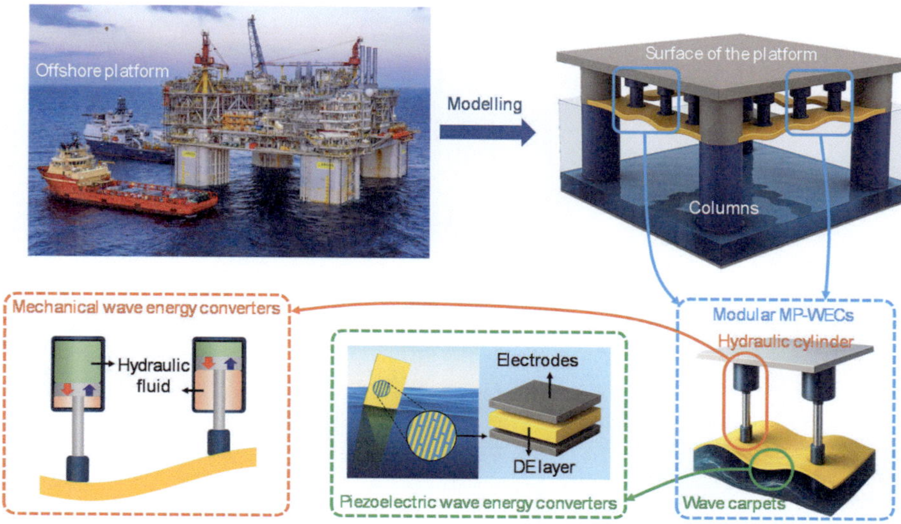


**Fig. 6.5** Illustration of how to provide clean energy for offshore platforms using a hybrid mechanical–piezoelectric WEC system

This integration harnesses the synergies between renewables and aquaculture provide on-site energy for aquaculture needs such as water pumping, feeding systems, and monitoring—replacing diesel generators that contribute to pollution. Environmentally, these hybrids can enhance biodiversity; for instance, wind turbine foundations or wave energy devices create artificial reefs that attract marine life, boosting fish stocks around aquaculture sites. Challenges like spatial conflicts and harsh weather are addressed through innovative designs, but the rewards include stabilized energy output—wave and tidal power complement intermittent wind and solar—and diversified revenue streams from energy sales and seafood harvests. As global policies push for net-zero emissions and sustainable blue economies, these combined systems represent a transformative approach, turning potential competitors for ocean real estate into collaborative powerhouses.

Key concepts in these hybrid systems revolve around multi-use platforms (MUPs), which are floating or fixed structures that co-locate energy generation with aquaculture, as shown in Fig. 6.6. One design features semi-submersible platforms where offshore wind turbines tower above, while sub-surface cages farm fish or shellfish below, utilizing the turbine's grid connection for power export and the structure's shade to regulate water temperatures for optimal growth. Another innovative approach integrates wave energy converters with seaweed farms, where the converters' motion dampens waves, creating calmer zones for macroalgae cultivation that absorbs $CO_2$ and provides biofuels or feed. Tidal energy systems, using underwater turbines akin to windmills, can pair with shellfish beds, as the steady currents enhance nutrient flow for faster mollusk growth, while solar panels mounted on floating platforms add a layer of photovoltaic power, providing energy

**Fig. 6.6** The integration of aquaculture with ocean-based renewable energy systems

as well as a shield to cool down the water underneath, valuable for aquaculture in hot climate. These designs often incorporate modular components for scalability, advanced sensors for real-time monitoring, and biofouling-resistant materials to withstand saltwater corrosion, ensuring long-term viability in dynamic marine settings.

Around the world, pioneering projects illustrate these concepts in action. The EU UNITED Project exemplifies co-location: in Germany, blue mussel and seaweed farms integrate with offshore wind farms for shared maintenance; in the Netherlands, floating solar pairs with seaweed; and in Belgium, flat oysters thrive amid wind turbines 30 km offshore. Australia's Carnegie Clean Energy's MoorPower system adapts CETO wave buoys to power salmon farms off Western Australia. Meanwhile, the Wave Dragon and Seaweed Energy Solutions partnership in Denmark developed a concept that merges wave energy with aquaculture, where overtopping waves generate power while creating sheltered areas for seaweed.

## 6.3 Digitalization, Monitoring, Management and Maintenance

The expansion of offshore renewable energy—encompassing wind, wave, tidal, and floating solar technologies—requires robust strategies for ensuring continuous, efficient, and safe operation. These systems must withstand some of the harshest conditions on Earth:

saltwater corrosion, high winds, extreme wave forces, marine growth, and limited physical access. To reduce maintenance costs, maximize energy output, and extend the lifespan of equipment, developers and operators are now relying heavily on integrated digital technologies. At the core of this transition are systems based on real-time sensing, cloud-based data analytics, predictive digital twins, automated tuning algorithms, and robotics-enabled intervention. These technologies allow for proactive monitoring, self-optimization, and reduced human dependency, especially for assets far offshore.

### 6.3.1 Embedded Sensors

Currently, offshore sensor technologies face their own formidable challenges. Traditional approaches often rely on bulky, bolt-on sensor modules designed for industrial settings on land. When transposed offshore, these designs reveal their weaknesses. Separate sensor packages, encased in rigid housings, can protrude awkwardly from renewable energy systems. Rather than blending into the structure, they add drag, invite marine growth, and in some cases even create new points of structural stress. The ocean, with its relentless cycles of wind, wave, and salt spray, present destructive power to such sensors.

Bulky sensors mounted externally are among the first to fail. Even when sensors survive, their signals may become unreliable as water ingress or micro-cracks interfere with measurements. In wave and tidal converters, where hydrodynamic efficiency is everything, the addition of traditional sensor housings can negatively impact the hydrodynamics thus lower the energy efficiency.

This is where embedded sensors offer a crucial shift. Instead of attaching bulky devices after the fact, sensors can be woven directly into the fabric of the system itself—bonded into turbine blades, laminated into wave carpets, or moulded into tidal foils. By becoming part of the material, rather than an add-on, they eliminate drag and reduce weak points. Strain gauges can be printed onto composite layers, accelerometers miniaturised into chip-scale devices, and load cells integrated into mooring cables as continuous fibres. These embedded designs not only survive longer in the marine environment but also capture more accurate, more localised data, reflecting the status of each key component of a structure.

Their modularity is another advantage. Offshore renewable systems increasingly rely on modular components—blades that can be swapped, pontoons that can be added, or tidal rotors assembled like building blocks. Embedded sensors fit naturally into this architecture. Each module carries its own intelligence, able to report independently on its health and performance. This decentralised approach prevents the system from being crippled by a single sensor failure and allows operators to replace or upgrade modules without overhauling the entire sensor network.

### 6.3.2 Digital Twins: Offshore-Cloud-Onshore

The offshore environment is rich in energy but poor in connectivity. Saltwater disrupt signals, and bandwidth is limited compared to terrestrial infrastructure. Yet without a reliable communication link, the immense intelligence gathered by embedded sensors risks being stranded at sea. That is why the second layer of intelligence in offshore renewable systems is all about data transmission—from offshore platforms to the cloud.

Modern systems overcome these challenges through a layered communication strategy. Optical fiber, often embedded in export cables that carry electricity back to shore, provides the gold standard of bandwidth and stability. Yet fiber is costly, and not all devices are connected. For these, satellite communications and offshore LTE/5G modems fill the gap. Because these channels are expensive and sometimes unstable, edge processors mounted on turbines or floating platforms act as intelligent gatekeepers. They collect high-frequency raw data from embedded sensors, run local analytics, and filter out noise. Only the filtered data are transmitted to shore. This reduces latency, preserves bandwidth, and ensures that decision-making software receives data that can be usable.

Onshore operators engage with digital twins—virtual replicas of the physical machines at sea. These twins do more than visualise; they simulate. Platforms such as Siemens Gamesa's digital operations suite, DNV's WindGEMINI, or Microsoft's Azure Digital Twins create dynamic models of turbines, cables, converters, and moorings, constantly updated with incoming offshore data. At the Hywind Scotland floating wind farm, Equinor uses digital twins to monitor mooring line tensions and hull dynamics, offering operators a predictive window into the future behaviour of the floating structures. If stresses begin to accumulate in a mooring chain, the system can forecast when and how failure might occur, prompting preventive maintenance before critical failure.

The integration of machine learning takes this predictive capacity even further. By training on a huge amount of operational data, digital twins can estimate the remaining useful life (RUL) of components, flag emerging risks, and even suggest operational adjustments. The AI models also help derive and provide information to locations that not directly covered by sensors.

### 6.3.3 Intelligent Management

The true power of offshore renewable energy systems lies not just in their ability to harvest energy from the elements, but in their capacity to think, adapt, and respond. Enabled by embedded sensors, offshore-to-cloud data pipelines, and digital twins, these systems are evolving from passive machines into intelligent, self-managing infrastructures. Where earlier generations of turbines or wave devices relied on fixed operating parameters, today's smart systems continuously reshape their behaviour in response to live conditions, squeezing more energy from every gust of wind or swell of the ocean.

In offshore wind, this intelligence is most visible in active control of yaw, pitch, and torque. Sensors feed wind speed/direction into algorithms that adjust blade angles and rotational speeds. Crucially, these adjustments are not just about chasing marginal energy gains; they are also about protecting the machine. By carefully balancing output with fatigue loading, intelligent control extends component lifetimes and reduces the frequency of costly offshore maintenance missions.

Wave and tidal energy systems push this principle even further, because the forces they face are nonlinear, multi-directional, and irregular. To capture energy efficiently and mitigate emtreme loads, converters cannot remain rigid—they must adapt the sea. Companies like CorPower Ocean have pioneered phase-control algorithms that keep buoyant converters "in tune" with the rhythm of incoming waves, adjusting internal pressures and PTO settings to maintain resonance. Other devices deploy adaptive damping systems or adjustable flaps that respond instantly to wave period, height, and direction. By letting machines flex and adapt rather than resist, these smart controls can boost efficiency compared to static designs, while also ensuring survival in the harshest storms.

Even offshore solar platforms are becoming intelligent organisms. Smart systems control mechanical components to tilt or rotate panels to track the sun's path, while in strong winds the same systems flatten or reorient the modules to reduce drag and prevent damage. By linking these controls with weather forecasting data, floating solar farms can act preemptively: folding arrays before storms, then autonomously resuming optimal orientation once the skies clear. This coupling of foresight and automation can potentially transform marine photovoltaics from vulnerable experiments into robust, utility-scale assets.

One of the most transformative promises of embedded intelligence in offshore renewables lies in the concept of fatigue life. Every blade, cable, or foundation in a marine energy system is constantly flexing, bending, or vibrating under wind and wave loading. Embedded sensors translate these structural responses into measurable fatigue signals—tiny changes in strain, vibration, or acoustic emissions that accumulate over time. Advanced structural algorithms then interpret these signals to estimate how much useful life remains before a crack propagates, a bolt loosens, or a coating fails. In effect, the structure itself becomes a living diary of wear and tear, continuously reporting on its health. Combined with digital twins, this means operators can move beyond scheduled maintenance calendars to a condition-based approach: servicing assets when they need it.

When a fault is detected or predicted by digital twin analytics, robotic maintenance systems increasingly form the frontline response. Beneath the waves, Remotely Operated Vehicles (ROVs) and Autonomous Underwater Vehicles (AUVs) have become indispensable for servicing turbine foundations, mooring chains, and subsea cables, including the removal of biofouling.

Above the surface, aerial drones can apply protective coatings, conduct welding, or deliver adhesive patches.

# 7 Epilogue

This book began with a question: how can humanity turn the ocean—so vast, so restless, so unpredictable—into a reliable partner in the transition to clean energy? The chapters that followed traced an arc from first principles to futuristic visions, each section building a layer of understanding about the technologies, challenges, and opportunities that define offshore renewables.

The first lesson is clear: offshore renewables are no longer science experiments. They are industrial realities. Offshore wind alone has moved from small demonstration projects to gigawatt-scale farms that power millions of homes. Wave and tidal technologies, though slower, are progressing from prototypes to pre-commercial deployments. Floating solar, a newcomer, is already beginning to challenge high seas. Together, these industries mark a turning point: the ocean is becoming a central pillar of the global energy system.

The second lesson is that technology alone is not enough. What makes offshore renewables viable is the fusion of machines with intelligence. Embedded sensors, digital twins, and robotic maintenance are not accessories; they are lifelines. Without them, offshore assets would remain fragile, unpredictable, and uneconomical. With them, they evolve into adaptive infrastructures capable of lasting decades.

The third lesson is about convergence. Offshore renewables are not isolated silos but hubs where multiple disciplines collide: material science meets hydrodynamics, robotics meets AI, energy economics meets marine ecology. This convergence is industrially profound: it creates new supply chains, new jobs, and new economies. Shipyards that once built oil rigs are now fabricating floating wind platforms. Robotics firms that serve civil purposes are pivoting to renewable inspection and maintenance. The transition of maritime power systems goes hand in hand with offshore hybrid energy platforms. The ocean is becoming not just an energy frontier but an ecosystem.



L. Huang, *Ocean Renewable Energy Systems*, Synthesis Lectures on Renewable Energy Technologies, https://doi.org/10.1007/978-3-032-10576-9_7

Looking forward, the possibilities expand beyond imagination: floating wind turbines linked with wave buoys, tidal turbines anchoring subsea networks, and solar platforms stretching across sheltered bays. Each unit is intelligent, self-monitoring, and self-maintaining, but together they form a vast, coordinated infrastructure. Electricity is not the only output: some platforms produce green hydrogen, others support desalination, and still others host aquaculture or ocean monitoring equipment.

This vision is not without challenges. The ocean will never be fully tamed, and climate change promises to intensify storms and sea-level rise. The technologies described in this book will not eliminate risk, but they will enable us to live with it wisely.

Ultimately, the greatest transformation may not be technological but philosophical. For centuries, humanity has treated the ocean as a void to be crossed or a storehouse to be exploited. Offshore renewables invite us to see it differently: as a partner, a collaborator, a dynamic environment in which we can learn and obtain far more than what we need on lands.

The task before us is to build that future with humility and imagination. The ocean will test us, as it always has. But if we succeed, the generations to come will inherit more than clean and affordable power. They will inherit a world where technology and the sea move together as partners in renewal.

# Correction to: Introduction

**Correction to:**
**Chapter 1 in: L. Huang, *Ocean Renewable Energy Systems*, Synthesis Lectures on Renewable Energy Technologies, https://doi.org/10.1007/978-3-032-10576-9_1**

The original version of Chapter 1 was inadvertently published with an incorrect reference list which has now been rectified. The chapter has been updated with the changes.

The updated version of this chapter can be found at
https://doi.org/10.1007/978-3-032-10576-9_1



L. Huang, *Ocean Renewable Energy Systems*, Synthesis Lectures on Renewable Energy Technologies, https://doi.org/10.1007/978-3-032-10576-9_8

MIX
Papier aus verantwortungsvollen Quellen
Paper from responsible sources
FSC® C105338

If you have any concerns about our products,
you can contact us on
**ProductSafety@springernature.com**

In case Publisher is established outside the EU,
the EU authorized representative is:
**Springer Nature Customer Service Center GmbH**
**Europaplatz 3, 69115 Heidelberg, Germany**

Printed by Libri Plureos GmbH
in Hamburg, Germany